FLIGHT WAYS

Critical Perspectives on Animals: Theory, Culture, Science, and Law

Critical Perspectives on Animals: Theory, Culture, Science, and Law
Series Editors: Gary L. Francione and Gary Steiner

The emerging interdisciplinary field of animal studies seeks to shed light on the nature of animal experience and the moral status of animals in ways that overcome the limitations of traditional approaches to animals. Recent work on animals has been characterized by an increasing recognition of the importance of crossing disciplinary boundaries and exploring the affinities as well as the differences among the approaches of fields such as philosophy, law, sociology, political theory, ethology, and literary studies to questions pertaining to animals. This recognition has brought with it an openness to a rethinking of the very terms of critical inquiry and of traditional assumptions about human being and its relationship to the animal world. The books published in this series seek to contribute to contemporary reflections on the basic terms and methods of critical inquiry, to do so by focusing on fundamental questions arising out of the relationships and confrontations between humans and nonhuman animals, and ultimately to enrich our appreciation of the nature and ethical significance of nonhuman animals by providing a forum for the interdisciplinary exploration of questions and problems that have traditionally been confined within narrowly circumscribed disciplinary boundaries.

The Animal Rights Debate: Abolition or Regulation?, Gary L. Francione and Robert Garner
Animal Rights Without Liberation: Applied Ethics and Human Obligations, Alasdair Cochrane
Experiencing Animal Minds: An Anthology of Animal-Human Encounters, edited by Julie A. Smith and Robert W. Mitchell
Animalia American: Animal Representations and Biopolitical Subjectivity, Colleen Glenney Boggs
Animal Oppression and Human Violence: Domesecration, Capitalism, and Global Conflict, David A. Nibert
Animals and the Limits of Postmodernism, Gary Steiner
Being Animal: Beasts and Boundaries in Nature Ethics, Anna L. Peterson

FLIGHT WAYS

Life and Loss at the Edge of Extinction

Thom van Dooren

Columbia University Press
New York

Columbia University Press
Publishers Since 1893
New York Chichester, West Sussex
cup.columbia.edu

Library of Congress Cataloging-in-Publication Data
Van Dooren, Thom, 1980–
Flight ways : life and loss at the edge of extinction / Thom van Dooren.
pages cm. — (Critical perspectives on animals. Theory, culture, science, and law)
Includes bibliographical references and index.
ISBN 978-0-231-16618-8 (cloth : alk. paper)
ISBN 978-0-231-53744-5 (e-book)
1. Birds—Extinction. I. Title.
QL677.4.V36 2014
598.13′8—dc23
2013044351

∞

Columbia University Press books are printed on permanent and durable acid-free paper.
This book is printed on paper with recycled content.
Printed in the United States of America
c 10 9 8 7 6 5 4 3 2 1

COVER PHOTO: © Wayne Kryduba/Mira.com
COVER DESIGN: Mary Ann Smith

For my parents

who taught me a profound sense of wonder and

an abiding respect for our living world

CONTENTS

ACKNOWLEDGMENTS

Many people have contributed to the writing of this book. I am grateful to all of those who have commented on draft chapters and presentations, as well as to everyone who has agreed to share their views and expertise with me in the many interviews and discussions that I draw on here. In particular, this book has benefited from an ongoing collaboration with Deborah Bird Rose, whose scholarship and friendship continue to inspire me. Donna Haraway also provided generous and insightful feedback on the draft chapters. I remain profoundly indebted to both of these remarkable scholars, who have, through their written work and personal guidance, shaped so much of my thinking and writing.

Among the many other people who have contributed significantly to this book, I would like to offer particular thanks to Matt Chrulew, Michelle Bastian, Emily O'Gorman, Jeff Bussolini, Eben Kirksey, Jodi Frawley, Heather Goodall, Jake Metcalf, Anna Tsing, Marc Bekoff, Maria Puig de la Bellacasa, Jim Hatley, and Rick De Vos. While many biologists and conservationists met with me to talk about birds, a few went above and beyond to provide me with their insights and, in some cases, access to restricted areas. In this regard, I am particularly thankful to Rich Switzer, Hob Osterlund, John French, John Marzluff, Alan Lieberman, and Chris Challies.

I would also like to thank the members of the academic communities that have been my home at various periods during the writing of this book. I began this research while I was a Chancellor's Postdoctoral Fellow at the University of Technology Sydney and completed it after moving to the Environmental Humanities program at the University of New South Wales. Both institutions provided a stimulating academic environment along with financial support for fieldwork. UTS also covered the cost of a four-month visiting position at the University of California, Santa Cruz, which provided me with yet another set of stimulating interlocutors. The vast majority of the funding for the fieldwork behind this book was provided by the Australian Research Council, in the form of a joint grant with Deborah Bird Rose focused on extinction in the broad Pacific Ocean region (DP110102886).

Finally, I would like to thank the birds themselves. Although they were not informed or perhaps even willing participants in the research—in fact, many of them were and remain captive—in a host of different ways they inspired me to try to say a little more about their precarious ways of life.

FLIGHT WAYS

INTRODUCTION

Telling Lively Stories at the Edge of Extinction

How else could a book about birds and extinction begin, but with the tragic story of the Dodo? In death, this bird from a small island in the western Indian Ocean has taken on a strange celebrity, becoming something of a "poster child" for extinction. And yet, many of the specific images and ideas about the Dodo that circulate in people's imaginations are highly speculative. Ultimately, a great deal remains unclear about what kind of a bird the Dodo was, how it lived, and when it passed from the world. While reports, sketches, and paintings of the Dodo survive from the seventeenth century, it is difficult to determine which of them is accurate and based on firsthand experience. Like a game of telephone, or Chinese whispers, it seems that many of these accounts and images were themselves based on other accounts and images, alongside varying degrees of poetic license (Hume 2006).

What we do know, however, is that Dodos (*Raphus cucullatus*) were large, flightless birds who made their homes exclusively on the island of Mauritius.[1] They probably ate mostly the fallen fruit available to a ground-dwelling bird, along with some seeds, bulbs, crustaceans, and insects. Fruit would have been abundant on the island prior to human arrival, when there were also no other terrestrial mammals present (Livezey 1993:271). In the absence of these mammals, Dodos likely had fewer competitors for these foods than did birds in many other places, but importantly, they also had no significant predators themselves—a situation that did not prepare them well at all for what was to come with the arrival of humans.

It is unclear who the first people to set eyes on the peculiar form of the Dodo were. Perhaps they were among the Arab traders who likely discovered the island in the thirteenth century. Or perhaps they were Portuguese sailors, among those who started visiting the island a few hundred years later (from 1507). As far as is known, however, neither of these groups settled on Mauritius, and no documentary evidence of an encounter with a Dodo remains.

The first reliable accounts of the Dodo were written by the Dutch after they arrived on the island in 1598 (Hume 2006:67). For roughly the next century, the Dutch East India Company used Mauritius as a "pasturing and breeding ground for livestock and a source of wild native meat" (Quammen 1996:265). This was the beginning of the end for the Dodo. Not only were they themselves on the menu—along with tortoises and a number of other local birds—but the various mammals that were intentionally and accidentally introduced to the island by the Dutch took their own huge toll.

Part of the problem for Dodos was undoubtedly their susceptibility to capture by hungry sailors and settlers. As flightless birds who had no previous experience of predators, they were easily captured by hand or beaten with a stick (Quammen 1996:266–68). While there have been frequent suggestions over the past few hundred years that Dodo meat was very unpalatable and infrequently consumed, that does not seem to have been the case. Paleontologist and Dodo expert Julian Hume (2006:80) has provided details of numerous firsthand accounts of the Dutch "relishing" the meat—in particular, the breast and stomach—and daily catching and eating many of these birds.[2]

It is likely, however, that the biggest problems that the Dodo faced after the arrival of humans on Mauritius were the other species of animals that came along on the journey. Foremost among them, chronologically at least, was probably the black rat (*Rattus rattus*). As in so many other places that European ships docked in the period, rats arrived early and with devastating force. Dodo eggs and young chicks, which up until this time would have required little protection, were an easy source of food. A little later, in the first decades of the seventeenth century, other new species joined them—notably, crab-eating macaques, goats, cattle, pigs, and deer. All these animals likely played a role in the decline of the Dodo: as predators, competitors for food, or both (Hume 2006:83).

No visitors to the island recount seeing a Dodo after the 1680s, perhaps a little earlier, and all evidence suggests that the species was extinct by the end of the seventeenth century (Hume, Martill, and Dewdney 2004). After thousands of years of peacefully gorging on fruits, the Dodo was suddenly thrust into an encounter with European culture, and just as quickly slipped out of the world.

While this was by no means the first species in whose loss humans were centrally involved, the Dodo inhabits a peculiar and iconic place in many contemporary accounts of extinction. This bird, and this biological process, have become strangely synonymous. If you ask the next person you see what they know about the Dodo, you might be told that it lived in Mauritius; you might even be told that it was a flightless bird; you will definitely be told that it is extinct.

"Dead as a Dodo"; little else about these birds seems to linger in our imaginations.

Perhaps this is because so little else is known with certainty about the species. But perhaps another reason for this close association between the Dodo and extinction is the particular way that this bird entered into written history. According to Beverly Stearns and Stephen Stearns (1999), the Dodo has the dubious honor of being "the first species whose extinction was conceded—in writing—to have been caused by humans" (17; see also Quammen 1996:277).[3]

I can offer no guarantee that the Dodo was actually *the* first species to be written about in this way, but it was certainly among the first. This was an extinction that occurred in the midst of the emergence of a slow realization by some European explorers and colonists that they might have huge impacts on the environments they were visiting, especially those of small islands. As environmental historian Richard Grove has noted, Mauritius was cited at the time as a key example of this potential. As forests were cleared and animal and mineral resources depleted, a "coherent awareness of the ecological impact of capitalism and colonial rule began to emerge" (Grove 1992:42). On Mauritius, however, it was too little, too late—both for the Dodo and for the numerous other species lost during this same period.

And so the Dodo entered into written accounts as a species driven to extinction by human activity, its fate strangely bound up with a dawning historical awareness that human activity might not just kill individual plants and animals, sometimes in their thousands, but also bring to an end whole ways of life. As a result of this awareness, the loss of species might be understood and narrated in a way that significantly *implicates* us—causally, perhaps emotionally, and certainly ethically. This is our sad inheritance from the Dodo.

In an important sense *Flight Ways: Life and Loss at the Edge of Extinction* is a continuation of the now well-established tradition of telling "extinction stories" that implicate people. But it is also an effort to tell these all-too-familiar stories in a new way. Specifically, the approach to thinking through extinction taken up in this book centers on "avian entanglements." Which is to say that this is a book about birds and their relationships, about the webs of interaction in which living beings emerge, are held in the world, and eventually die. Life and death do not take place in isolation from others; they are thoroughly relational affairs for fleshy, mortal creatures. And so it is, in the worlds of birds—woven into relationships with a diverse array of other species, including humans. These are relationships of co-evolution and ecological dependency. But they are also about more than "biology" in any narrow sense. It is inside these multispecies entanglements that learning and development take place, that social practices and cultures are formed. In short, these relationships produce the possibility of both life and any given way of life. And so these relationships matter. This is true at the best of times, but in times like these when so many species are slipping out of the world, these entanglements take on a new significance.

Flight Ways is composed of five extinction stories, each focused on a group of threatened birds. In emphasizing these birds' entanglements, the book draws us into a deeper understanding of who they are, who we are, and ultimately how it is that we all "become together" (Haraway 2008), for better or worse, in a shared world. Through this lens, it is clear that much more than is often appreciated is at stake in the disappearance of birds. And so we are able to understand in new ways the diverse significances of extinction: What is lost when a species, an evolutionary lineage, a way of life, passes from the world? What does this loss mean within the particular multispecies community in which it occurs: a community of humans and nonhumans, of the living and the dead? How might we think through the complex place of human life at this time: simultaneously, a/the central cause of these extinctions; an agent of conservation; and organisms, like any other, exposed to the precariousness of changing environments?

In focusing on entanglements, this book aims to present alternative understandings of extinction to those grounded in entrenched patterns of "human exceptionalism." This exceptionalism presents humans as fundamentally set apart from all other animals and the rest of the "natural" world (chaps. 2 and 5). In this context, extinction cannot help but be regarded as something that happens "over there" or out in "nature." In contrast, the approach taken in this book is grounded in an attentiveness to the diverse ways in which humans—as individuals, as communities, and as a species—are implicated in the lives of disappearing others. Paying attention to avian entanglements unsettles human exceptionalist frameworks, prompting new kinds of questions about what extinction teaches us, how it remakes us, and what it requires of us. This last question is of particular importance. Ultimately, this book is concerned with broad questions of ethics: What kinds of human–bird relationships are possible at the edge of extinction? What does it mean to care for a disappearing species? What obligations do we have to hold open space in the world for other living beings?

FROM DEEP WITHIN A TIME OF EXTINCTIONS

Sadly, extinction is not a topic that generates a great deal of popular interest at the present moment. I suspect, however, that in the future to come—if humanity is here at all—extinction will be among the handful of themes that is understood to be central, perhaps even definitional, of our time. We are the generations that are overseeing the loss of so much of the diversity of living forms on this planet, the generations that are perhaps yet to fully understand and respect the significance of the intimately entangled, co-evolved, forms of life with which we share this planet.

According to some biologist and paleontologists, this period may well be Earth's sixth mass extinction event (Kingsford et al. 2009); according to others, we are not quite there yet, but certainly on the way (Barnosky et al. 2011). Past mass extinction events, like the one that took the dinosaurs roughly 65 million years ago at the end of the Cretaceous and the even larger end-Permian event around 250 million years ago, saw losses of more than 75 percent of Earth's species (Jablonski and Chaloner 1994; Raup and Sepkoski 1982). In place of meteor impacts, volcanic eruptions, and the various other forms of massive upheaval proposed as possible causes for the previous "big five" events, it is tragically clear that ours is an *anthropogenic* extinction event. Current deaths of species are being brought about, directly and indirectly, by a range of interwoven human activities—including the destruction of habitat, the promulgation of introduced species, direct exploitation and hunting, the indiscriminate introduction of a range of new chemicals and toxins, and now increasingly the various impacts of climate change.[4]

The scale of this loss is unknown and unknowable with any real certainty. Biologist Richard Primack (1993) estimates that the current rate of extinction is likely 100 to 1,000 times greater than would be expected as a result of normal "background extinction."[5] According to some scientists, we are now on a trajectory to lose between one-third and two-thirds of all currently living species (Myers and Knoll 2001:5389). Within this broader space of loss, some taxonomic families will be hardest hit. Frogs, salamanders, and other amphibians, for example, are considered to be at particular risk, with approximately one-third of all species now thought to be endangered or recently extinct (Stuart et al. 2008).

Birds, too, have also been hard hit by extinction. In the past 500 years, 153 documented avian extinctions have occurred (Birdlife International 2008:4). It is likely, however, that the actual number is much higher, as some species that are listed as "critically endangered" are in actuality already extinct, and others will disappear without having been documented at all. Today, one in eight known bird species is thought to be threatened with global extinction, while among some taxonomic families, the number is much higher (Birdlife International 2008:5)—for example, 82 percent of all albatross species are threatened (chap. 1).

Those birds that make their homes on islands have also tended to fair particularly badly. While "only" 20 percent of the world's bird species are confined to islands, approximately 90 percent of the avian extinctions that have occurred in recorded history have been those of island inhabitants (Quammen 1996:264). For example, in and around the Pacific Ocean where much of this book is set, successive waves of human settlement (and colonization and occupation) have taken their toll (Steadman 2006). As biologist John Marzluff (2005) has simply put it: "In little over a thousand years we have extinguished more than half of all the bird species that occupied the lush islands of the tropical Pacific" (256). As we enter more deeply into this current period of loss, however, mainland birds—including some of those once thought exceedingly common—are also increasingly being placed at risk of extinction (for example, the Indian vultures discussed in chap. 2).

But despite all these known losses—from the Dodo to the Passenger Pigeon (*Ectopistes migratorius*) and the King Island Emu (*Dromaius ater*)—our knowledge of this situation remains thoroughly partial. The total number of species being driven over the edge in this "time of extinctions" (Rose and van Dooren 2011) simply overwhelms our capacity for understanding. We just do not know how many are being lost: How could we, when we do not even know how many species there are on this planet with any reasonable degree of certainty? While we sometimes hear about a handful of charismatic endangered species, countless others go completely unremarked on and even unnoticed (at least by modern science, and perhaps humans more generally).[6] As biologist Bruce Wilcox (1988) notes, "[F]or every species listed as endangered or extinct at least a hundred more will probably disappear unrecorded" (ix).

TELLING LIVELY STORIES ABOUT EXTINCTION

Flight Ways is set within the shadow of this incredible loss. It is in this context that it asks about the nature of extinction and why and how it matters. As a whole, this book is grounded in the conviction that there is no single "extinction" phenomenon. Rather, in each case there is a *distinct* unraveling of ways of life, a distinctive loss and set of changes and challenges that require situated and case-specific attention. In delving into the lives and deaths of particular bird species, this book attempts to draw out their "entangled significance." Across simultaneously "biological" and "cultural" domains, the book explores some of the ways in which diverse living beings—humans and not—are drawn into the extinctions of others. Far more than "biodiversity"—at least in the narrow sense that the term is often used—is at stake in extinction: human and more-than-human ways of life, languages, ways of mourning and being with others, even livelihoods and diverse cultural and religious worlds are often drawn into the fray as species move toward, and then beyond, the edge of extinction.

Narrative is my way into this complexity; stories allow us to hold open simultaneously a range of points of view, interpretations, temporalities, and possibilities (Griffiths 2007). But this book takes a particularly "lively" approach to telling stories about life and death in the shadow of extinction.[7] It is an effort to weave tales that add flesh to the bones of the dead and dying, that give them some vitality, presence, perhaps "thickness" on the page and in the minds and lives of readers. This is an inherently multidisciplinary task, and so the stories that I tell in this book engage with the literatures of biology, ecology, and ethology (the study of animal behavior and cognition), as well as with interviews and conversations with scientists of various kinds. In drawing on the natural sciences, I hope to invite readers into a sense of curiosity about the intimate particularities of these disappearing others: how they hunt or reproduce, how they take care of their young or grieve for their dead, how they make themselves at home in the vast Pacific Ocean or along an urban coastline. Paying attention to the details of how these lives are, or once were, lived invites us into a sense of wonder.

Rendered in this way, these creatures become more than a name—no longer an abstract Latin binomial on a long list of threatened species, but a complex and precious *way of life*. And so this approach to storytelling is a core part of my effort to capture a fuller sense of what extinction *is* and to insist that nonhuman others are not simply "life forms," but "forms of life" (Helmreich 2009:6–9). I draw this distinction from the anthropologist Stefan Helmreich (2009), who puts it to productive use to explore the entanglement of various "life forms"—understood as organisms in ecological relationship—with diverse "forms of life," which, adapting Ludwig Wittgenstein, he understands as "those cultural, social, symbolic, and pragmatic ways of thinking and acting that organize human communities" (6). There is, however, no reason why a line must be drawn at the human, and so the stories in this book are particularly interested in the "forms of life" that have emerged, and are possible, for some of the many disappearing other-than-human "life forms" that populate this planet. As will be discussed in detail in chapter 1, this understanding of birds (and other organisms) as life forms *with* a form or way of life is central to my notion of species as "flight ways."

In drawing on the perspectives of the natural sciences in taking up this topic, my intention is not to imply that they offer us the only—or even necessarily the best—means of understanding the lives and deaths of birds. And yet, some of the work within these disciplines has provided ways of knowing that deeply influence my own appreciation of the world and my sense of the significance of extinction. As such, I draw on work in the natural sciences that I think helps to animate a fuller and richer sense of the lives of particular beings. This approach takes seriously Donna Haraway's (2008) injunction to practice a genuine curiosity in our philosophical engagements with a more-than-human world; it is a practice grounded in "knowing more at the end of the day than at the beginning" (36).

As I researched each chapter—reading, thinking, and conducting interviews and fieldwork—I got to know these species in new ways. In each case, I was surprised by the way in which "knowing more" draws us into new kinds of relationships and, as a result, new accountabilities to others. As I came to understand a little better the particular dynamics of the relationship between Little Penguins and the coastlines that they nest on, I began to appreciate in new ways the ethical weight of our destructive actions in these places (chap. 3). As I reflected on the complex ecological and social relationships that Hawaiian Crows live within, I also developed a new awareness of the significance of their disappearance from island forests (chap. 5). And so I came to appreciate the ethical work that these stories may do in the simple act of making disappearing others thick on the page, exposing readers to their lives and deaths in a way that might give rise to genuine care and concern.

My guide in thinking through the ethics of storytelling in this way is James Hatley's work on narrative and testimony in the face of the Shoah. Hatley forcefully reminds us of the ethical demands of the act of writing: of giving an account or telling stories. In place of an approach that would reduce others to mere names or numbers, in place of an approach that aims for an impartial or "objective" recitation of the "facts," Hatley argues for a form of witnessing that is from the outset already seized, already claimed, by an obligation to those whose stories we are attempting to tell. This is particularly the case when our stories play the role of witness or testimony to the suffering and deaths of others (Hatley 2000:114). In the context of extinction, these kinds of stories are not an attempt to obscure the truth of the situation, but to insist on a truth that is not reducible to populations and data: a fleshier, more lively, truth that in its telling might draw us all into a greater sense of accountability (van Dooren 2010; but see Smith 2001:368). As William Cronon (1992) simply puts it: "Good stories make us care" (1374).

Consequently, at the same time as they may offer an account of existing relationships, stories can also connect us to others in new ways. Stories are always more than simply descriptive: we live by stories, and so they are inevitably powerful contributors to the shaping of our shared world. This is an understanding that works against any neat or straightforward division between the "real" and the "narrated" world (Kearney 2002:133–34). Instead, I see storytelling as a dynamic act of "storying" the world, utterly inseparable from lived experience and a vital contributor to the emergence of "what is." Stories arise from the world, and they are at home in the world. As Haraway (forthcoming) notes, "'World' is a verb," and so stories are "*of* the world, not *in* the world. Worlds are not containers, they're patternings, risky co-makings, speculative fabulations." Even a story that aims to be purely mimetic can never simply be a passive mirror held up to "reality." Stories are a part of the world, and so they participate in its becoming. As a result, telling stories has consequences: one of which is that we will inevitably be drawn into new connections, and with them new accountabilities and obligations.

And so the bird stories that this book tells/does are "lively" in both their message and their form—that is, in their commitment to the continuity of diverse ways of life, and in their attempt to enact stories as interventions into existing patterns of living and dying in an effort to work toward better worlds.

THE EDGES OF EXTINCTION

As previously noted, this book is animated and guided by a desire to weave stories that explore and convey the entangled significance of extinction. In so doing, a key part of my interest is in broadening our notion of what extinction actually *is*, beyond the simple black-and-white versions of it that often dominate. These conventional understandings center on the death of the last individual of a kind. We may not very often be sure if any given individual really is the last, but we are usually confident that if we did (or could) know for certain, then we would be able to pinpoint the precise moment of an extinction. The death of Martha the Passenger Pigeon at the Cincinnati Zoo in 1914, or that of an unnamed Po'ouli (*Melamprosops phaeosoma* [a Hawaiian honeycreeper]) in conservationists' hands in 2003, were in all likelihood simultaneously deaths of individuals and "extinctions" in this sense.

There is, of course, something entirely accurate about this understanding. Something important and profound took place with the deaths of these last individuals. And yet, the immensity and significance of extinction cannot be captured within these singular events, as though a species might be deemed to be extinct or not solely on the basis of the presence in the world of at least one individual of that kind/lineage. This understanding reduces species to specimens—reified representatives of a "type" in a museum of life—in a way that fails to acknowledge their entangled complexity (chaps. 1 and 2, in particular). The nomadic form of life of Passenger Pigeons, moving through the sky in flocks of hundreds of millions of birds that blocked out the sun, had long since come to an end when Martha passed away in 1914. As Passenger Pigeon numbers dwindled, the social and behavioral diversity of this unique way of life—of what it was to *be* a Passenger Pigeon in some fundamental sense—would also have broken down. Similarly, over the decades before Martha's death, the interspecies relationships that the Passenger Pigeon evolved and lived within would also have become increasingly fractured as these birds stopped playing any significant role in the lives and nourishment of diverse humans and nonhumans.[8]

A singular focus on Martha's death covers over all of this; it presents a species as somehow "ongoing" because one individual continues to draw breath in a zoo, while the entangled relations that in a nontrivial sense *are* this particular life form and its form of life, have long ago become frayed and disconnected.

The point here is not that a bird in a zoo is not a bird at all. Clearly, many birds are capable of living in a range of environments, of adapting to changed conditions: a species is not a single, narrow, and unchanging way of life—as is indicated so well by the numerous birds and other animals who have taken up residence within, sometimes as an integral part of the emergence of, "human" cities (Hinchliffe and Whatmore 2006; van Dooren and Rose 2012; Wolch 2002). Rather, the point is that the loss, the change and disruption—often accompanied by violence and suffering—that occurs in extinction must not be reduced to this one event. Instead, the deaths of these last individuals must be understood as singular losses in the midst of the tangled and ongoing patterns of loss that an extinction *is*.

This understanding of extinction is, of course, grounded in an attentiveness to entanglements. When species are understood as vast intergenerational lineages, interwoven in rich patterns of co-becoming with others (chap. 1), then their departure from the world cannot help but be felt in a range of complex and drawn-out ways. In an effort to take these entanglements seriously, this book focuses on some of the various "edges of extinction." In spending time in this terrain of living and dying, I have become acutely aware that extinction is never a sharp, singular event—something that begins, rapidly takes place, and then is over and done with. Rather, the edge of extinction is more often a "dull" one: a slow unraveling of intimately entangled ways of life that begins long before the death of the last individual and continues to ripple forward long afterward, drawing in living beings in a range of different ways (chap. 2, in particular).

As becomes clear in this book, these spaces at the edge of extinction are far from uniform. Each of the birds discussed draws us into a different set of relationships. In one case, it is a space in which countless albatross chicks die each year through the consumption of plastics and other toxins. In another, it is a space of contestation between penguins returning faithfully to a disappearing coastline that was once their nesting site and the people, dogs, and others that now also call this place home. In the context of Hawaii's crows, it is a space of both potential and actual grief and mourning, in which the deaths of others might provide powerful opportunities to relearn our place in a shared world.

In many of these cases, the edge of extinction is now also deliberately flattened and drawn out by active human intervention to conserve disappearing species. Through these efforts, species are held in the world for decades more than they might otherwise have survived. In addition, therefore, to being spaces of suffering, death, and loss, these edges of extinction are now often also places of intense hope and dedicated care. Chapter 4, in particular, explores the way in which the edge of extinction might be flattened through conservation efforts—in this case, with a focus on the iconic Whooping Crane. Here, my particular interest is in the strange juxtaposition of violence and care, of coercion and hope, that characterizes the lives and deaths of captive cranes (and many other species) at the "dull edge of extinction."

In short, these edges of extinction are varied, complex, and conflicted spaces in which diverse relationships, diverse multispecies communities, emerge as possibilities of life and death for everyone—not just the "endangered"—are remade.

THE STRUCTURE OF THIS BOOK

The approach taken in *Flight Ways* is situated within ongoing discussions in two emerging fields of scholarship: animal studies and the environmental humanities. Both are thoroughly interdisciplinary fields where the humanities and social sciences are drawn into conversation with the natural sciences. This book aims to contribute to both areas of scholarship, but also to encourage the deepening of dialogue between them.

Each of the chapters might be read in isolation. On the surface, each of them tells a unique and largely self-contained story, with occasional references to related discussions in other chapters. However, my intention is for the book to be read in order and as a whole. In gentle but important ways—ways that will hopefully become clear as the reader moves through the book—each chapter builds on those that precede it, taking for granted both concepts and commitments that are fleshed out more fully in earlier chapters.

Chapter 1 explores the plight of the Black-footed Albatrosses (*Phoebastria nigripes*) and Laysan Albatrosses (*P. immutabilis*) of Midway Atoll in the remote North Pacific Ocean. The chapter takes up this topic through a focus on the difficult work of fledging young albatrosses (that is, raising them until they are ready for flight): the creation of a solid pair-bond between breeding birds, the laying and incubating of eggs, the months of movement back and forth between land and sea in search of food to satisfy hungry young chicks. Through this account, the chapter proposes a particular understanding of what a species is, an understanding that focuses on the time, energy, and labor that are required to keep successive

Arctic Circle
45°
30°
Tropic of Cancer
INDIA
Equator
ATLANTIC
OCEAN
INDIAN
OCEAN
Tropic of Capricorn
30°
45°
60°
SOUTHERN
OCEAN
Antarctic Circle

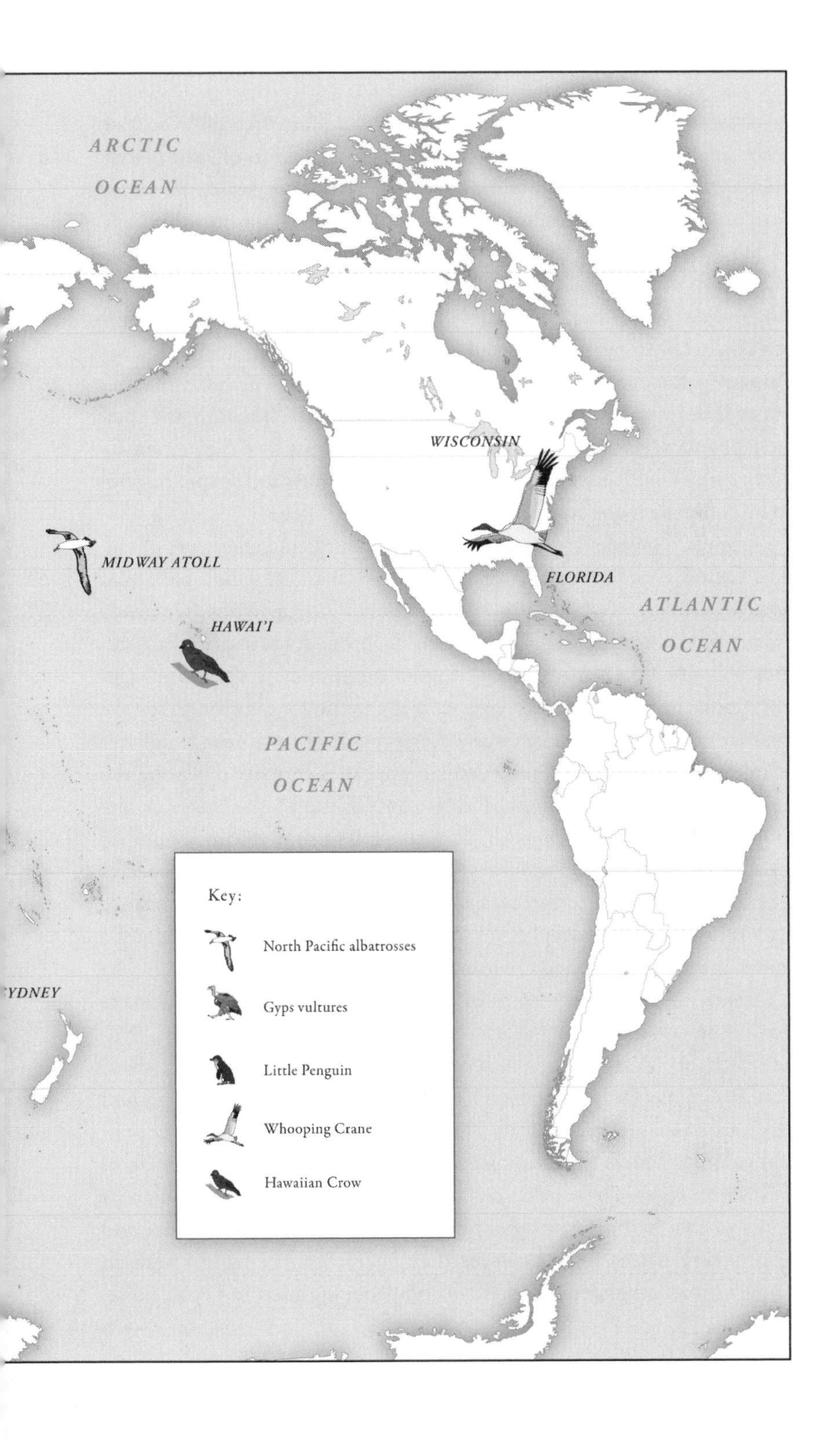
ARCTIC
OCEAN
WISCONSIN
MIDWAY ATOLL
FLORIDA
ATLANTIC
OCEAN
HAWAI'I
PACIFIC
OCEAN
YDNEY
Key:
North Pacific albatrosses
Gyps vultures
Little Penguin
Whooping Crane
Hawaiian Crow

generations in the world. In this context, species are incredible *achievements*: intergenerational lineages stretched across millions of years of evolutionary history. In our time, however, the circulating waste of human societies threatens the continuity of albatross species, harming and killing breeding birds and their young. In this context, the chapter focuses on the diverse temporalities enfolded at this site of encounter. Here, the daily lives of birds—and, ultimately, the futures of their species—come into contact with persistent pollutants and seemingly immortal plastics. Ultimately, the chapter explores some of the ways in which the difficult task of taking seriously these vastly different temporal horizons and their overlaps and intersections provides us with a fuller sense of the immensity of what is lost in extinction, while drawing us into new and deeper responsibilities for our living world.

Chapter 2 considers some of the contemporary entanglements of vultures (genus *Gyps*), people, cattle, and others in India, with a particular focus on the way in which lives and livelihoods are made possible inside interactions in a more-than-human world. In the context of Indian vultures, this situation is made more complex because these species are rapidly approaching extinction. When vultures are no longer around to take up the relationships that they once did, many other lives are made difficult or impossible—with poor and rural communities very often bearing the majority of the human burden. In this context, the chapter takes up the notion of the "dull edge of extinction" to explore some of the inequities of exposure to suffering that emerge inside relationships of multispecies dependency. This is a topic that can only take on increasing importance as we move ever more deeply into the current period of extinctions and a time of greater climatic and environmental change.

Chapter 3 takes up the story of a tiny colony of penguins that make their home just inside the mouth of one of Australia's busiest ports, Sydney Harbour. Members of the world's smallest penguin species, these Little Penguins (*Eudyptula minor*) stand roughly 1 foot (30 cm) tall and weigh around 2 pounds (1 kg). They also make up one of the last penguin colonies left on the Australian mainland and the last in the state of New South Wales. For roughly eight months of each year, these penguins return to this harbor, coming ashore at various places to lay eggs and fledge young. Increasingly, however, their burrows are being lost to them through urban development and its accompanying patterns of light, noise, and disturbance (in particular, predation by domestic dogs). This chapter explores the nature of these penguins' attachment to their specific breeding places, called "philopatry" or "site fidelity." Despite ongoing changes and increased danger, year after year they return. The chapter argues for an understanding of these breeding sites as "storied-places," invested with history and meaning for penguins. Consequently, it explores the ethical significance of destroying places that penguins (and others) are in an important sense tied to. The chapter asks: What kinds of ethical obligations might be opened up by a new sensitivity to the storying and place-making practices of penguins and other nonhumans?

Chapter 4 is focused on one of North America's longest-running conservation programs, that of the iconic Whooping Crane (*Grus americana*). For more than forty years, conservationists in the United States and Canada have worked to protect these birds and their wintering and summering grounds. On many levels, this is a story of care and success in which conservationist have managed to pull the species back from the edge of extinction—from fewer than 20 birds in the early twentieth century to roughly 600 today. This chapter takes up this conservation story through a close focus on the elaborate captive breeding and release program that for some young birds culminates in the use of ultralight aircraft to teach them a new migratory route. My particular interest is in the strange juxtaposition of care and violence that lies at the heart of this effort and the ethical dimensions of the human–crane relationships that are being established. Who suffers and who dies so that new populations of this species might make their way back into the world? On what grounds are the lives of some beings sacrificed for the sake of others, and might a concerted effort to inhabit and examine these complex and difficult situations—"staying with the trouble" (Haraway, forthcoming)—provide an opening into a more ethical mode of conservation?

Chapter 5 returns us to the heart of the Pacific Ocean, this time with a focus on the only endemic corvid species, the Hawaiian Crow (*Corvus hawaiiensis*). In 2002, the last free-living crow died. As forest-dwelling fruit specialists, these crows have been significantly affected by the degradation of local forests, as well as by increased predation and introduced diseases. This chapter considers the limited ethological literature on the ways in which crows (and corvids more generally) respond to the deaths of others of their kind. Much of the history of Western thought has utilized animals' understandings of and responses to death to construct a dualism between "the human" and "the animal." This dualistic thinking is at the core of a human exceptionalism that holds us apart from the rest of the world and, as such, contributes to our inability to be *affected* by the incredible loss of this period of extinctions, and so to mourn the ongoing deaths of species. In contrast to this tradition, this chapter explores some of the ways in which taking crows' grief seriously may, in fact, work to undermine our sense of human exceptionalism—in particular, by highlighting both a deep evolutionary continuity between humans and other social animals, and our ecological entanglement in a more-than-human world. In this way, telling stories about grieving crows may itself become an act of mourning extinctions. This would be a mode of mourning that does *not* announce the uniqueness of the human, but works to undo exceptionalism, drawing us into company with crows and others to grieve for the loss of a world that *includes us,* to grieve the countless deaths that constitute this time of extinctions.

Through each of these avian case studies, *Flight Ways* explores new modes of storytelling. Ultimately, it offers a call for stories, a call for new ways for figuring our place in and obligations to a rapidly changing world.

One

FLEDGING ALBATROSSES

Flight Ways and Wasted Generations

The places of albatrosses are beyond the inhabited limits of humanity, on spare, elemental islands that feel like the center of a waterbound planet. Yet humans touch them in all their haunts.

CARL SAFINA, "WINGS OF THE ALBATROSS"

In the middle of the North Pacific Ocean, at the far northwest end of the Hawaiian Archipelago, lie a few tiny coral and sand islands encircled by a small reef. These little patches of dry land in the midst of a vast expanse of water and sky are Midway Atoll. Roughly halfway between the United States and Japan, Midway is about as far as it is possible to be from the nearest continental landform and more than 1,200 miles (1,930 km) from the nearest significant human population, in Hawai'i (USFWS 2012). Each year, this place is home to a dizzying array of bird species who return to breed. Among this winged diversity are thousands upon thousands of breeding Laysan Albatrosses (*Phoebastria immutabilis*) and Black-footed Albatrosses (*P. nigripes*); this most isolated of locations—at least from a human perspective—is where these birds lay their eggs and watch over their young. And yet even here, life is profoundly influenced by far- reaching ripples of human activity.

In recent decades, one of the most visible anthropogenic impacts on many small Pacific islands like Midway has been the presence of plastic objects of all shapes and sizes. Circulating in the world's oceans in

A pair of breeding Laysan Albatrosses at Midway Atoll. (David Patte/U.S. Fish and Wildlife Service; CC BY 2.0)

ever-increasing quantities, these plastics pose significant risks to a wide range of marine animals who ingest or are entangled in them (Gregory 2009). The albatrosses of Midway are no exception. Heading skyward and seaward in search of food for growing chicks, albatrosses invariably collect plastic items that they mistake for food or that are entangled with favorite food items (like fish-egg clusters). These plastics are then delivered into the waiting mouths of hungry chicks, where they accumulate to contribute to malnourishment, dehydration, starvation, and various other health problems. While ingestion of plastic in this way is a problem the world over, the North Pacific is in many ways in a league of its own. It is now thought that Laysan Albatrosses, in particular, may "have a greater incidence, a wider variety, and larger volume of ingested plastic than any other seabird" (Auman, Ludwig, Giesy, et al. 1997:239; De Roy, Jones, and Fitter 2008).

At this time, however, the plastics and other toxic compounds circulating in the world's oceans and accumulating in albatross bodies threaten not only the lives of individual birds, but the future of their species, too.

Against this background of violence and potential loss, this chapter develops an understanding of species as "flight ways." This is an understanding that emphasizes the "embodied temporality" of species (Rose 2012b), asking us to pay attention to species as evolving "ways of life" that are shared, produced, and nurtured in the world through the work of successive generations of living beings. Thinking in this way requires us to work across entirely different temporal horizons: to think about species in a way that acknowledges that they are vast evolutionary lineages stretched across millions of years, while not losing sight of the fleeting and fragile individual birds whose lives and labors both constitute and enable the continuity of this larger species.

But these are not the only temporal frames at work. In addition, this chapter is concerned with the life spans of seemingly immortal plastics, the half-lives of persistent organic pollutants, the evolution of our own human species (the unwitting architect of the current period of mass extinction), and a range of other bodies and processes—all taking place across very different temporal horizons. In their messy coming together, with divergent and overlapping durations and possibilities, much is at stake—for albatrosses, but also for people and for the broader community of coevolved life to which we both belong.

Paying attention to these complex temporalities enables us to better understand the significance of the coming to an end of a way of life. But, importantly, it also draws us into new kinds of connections, and so new responsibilities (Bastian 2013), introducing us to an ethical claim that is made on us to hold open space in the world for other species. In this context, telling albatross stories is an effort to gather up disparate times and places—not to make them homogeneous and neat, but to ethically inhabit their complexity. In this time of incredible loss of species, we need stories that can travel far and fast—like an albatross riding the wind in search of food. At the very least, we need stories that can travel as well as plastics and other pollutants, both of which can make use of vast oceanic and atmospheric systems for their movements. We need stories that can reconnect people with the distant and ongoing impacts of their waste in a way that may make a difference for the future of generations of albatrosses and all those other species with whom they are entangled.

WANDERING THE OCEAN

Albatrosses are pelagic birds of the most extreme variety. In contrast to the many seabirds who spend much of their time along the coast, albatrosses live almost all their lives far out to sea.[1] Nomadic birds, often referred to as "wanderers," albatrosses comfortably cross huge expanses of water each day: "Big birds in big oceans, albatrosses lead big, sprawling lives across space and time, travelling to the limits of seemingly limitless seas" (Safina 2008:20). In a single day, an albatross will commonly cover several hundred miles of ocean, and many have been recorded travelling much farther.

While the speed at which they move is itself reasonably impressive—with averages of 30 to 50 miles per hour (50–80 kph) commonly recorded (Lindsey 2008:68–69)—it is the *sustained* manner in which albatrosses travel that is the key to their ability to wander far and wide. While many other birds, especially migratory species, can cover large distances in short periods of time, these are generally relatively infrequent trips. For albatrosses, though, this kind of travel takes place day in, day out (Lindsey 2008:69). On average, albatrosses are thought to spend roughly 95 percent of their lives at sea, and for the majority of this time they are in flight, gliding just above the ocean's shifting surface (Safina 2008:21).

In order to remain aloft for such extended periods of time without becoming exhausted, albatrosses flap their wings as little as possible and instead rely on the wind to do most of the work. With only minor adjustments to their body and wing orientation, albatrosses can soar virtually effortlessly across the waves in an up-and-down, seesawing movement (called "dynamic soaring").[2] As biologist Scott Shaffer (2008) has put it: "[T]he albatross body plan is perfectly designed to maximize gliding performance at high speeds with a minimum of effort" (153).

And yet, despite their adaptation to and immersion in a world of wind and waves, albatrosses remain utterly tied to the land as well, required to return each year to lay eggs and fledge young.[3] Midway Atoll is a particularly important breeding place for both Black-footed and Laysan Albatrosses, having by far the largest population of each species: more than one-third of the global population of Black-footed Albatrosses and over two-thirds of the world's Laysan Albatrosses breed at Midway. Pressed onto a few tiny, exposed, low-lying islands, on a single atoll, these colonies are vital to the long-term survival of both species.

For roughly eight months of each year, breeding albatrosses at Midway are engaged in a seemingly endless movement between land and water, alternating between egg or chick duties on land and days or weeks at sea in search of food. From when the single egg is laid in November or December, the breeding period is roughly broken down into three parts. It will take about two months for the egg to hatch, during which time the breeding pair take turns, one incubating the egg while the other goes to sea. The bird on land is often there for periods of more than twenty days at a time, steadily losing weight with no access to food or water. When the egg hatches, life only becomes more difficult. The chick must be brooded for several days, and then guarded constantly in its small and vulnerable state for roughly another month. Once it is big enough to be left on its own, both parents take to the sea in search of the huge quantities of food that are required to nourish a growing chick (Naughton, Romano, and Zimmerman 2007:3–4; Rice and Kenyon 1962).

For about the next five months, albatross parents travel far and wide collecting food. During this period, short trips into local waters are interspersed with two-week-long trips into distant, often subarctic, waters (Fernandez et al. 2001; Hyrenbach, Fernandez, and Anderson 2002). It is only by virtue of the ease with which they travel long distances that albatrosses are able to raise chicks in a warm tropical climate while drawing on the rich food offerings of distant cooler waters. By the time a chick is ready to fledge, it is roughly the same size as its parents. In terms of distance travelled, quantity of food collected, and duration of the breeding season, the effort involved in successfully fledging an albatross chick is simply massive. This is, as ornithologist Terence Lindsey (2008) has noted in relation to the larger Wandering Albatross (*Diomedea exulans*), "a herculean provisioning feat" (94).

It is perhaps these high demands that explain why Laysan and Black-footed Albatrosses do not attempt to breed as soon as they reach sexual maturity (at roughly the age of five). In fact, most do not breed for another three or four years, and some put it off considerably longer (Naughton, Romano, and Zimmerman 2007:3; Rice and Kenyon 1962:520–21). During these additional years, it is thought that juvenile albatrosses are gaining the knowledge and skills necessary to provide adequately for a growing chick (Lindsey 2008:82).

But in addition to demanding a lot of each of the individual parents, a very strong pair-bond between breeding birds is essential to successfully fledge albatrosses. Breeding is simply not something that can be carried out by a single bird at any stage in the process. If one parent dies or abandons the effort, the chick will not survive. When the partner of an incubating bird does not return to relieve it, the bird often stays with the egg for as long as possible—sometimes well over a month, to the point of its own starvation—but it will eventually be forced to abandon the egg and return to the sea in search of food (Rice and Kenyon 1962:545). Even once the egg is hatched and the chick does not have to be guarded, a single bird is simply unable to provide enough food for the chick to grow and fledge healthily.

It makes sense, therefore, that before a first breeding attempt, a great deal of time is invested "in forging uncommonly stable and intimate partnerships" (Lindsey 2008:82–83). While many juvenile albatrosses start returning to their natal colonies during breeding season when they are three to five years of age, they do not actually begin to breed for several more years. During this time, they busily search for a suitable mate. Each breeding season, when they return to Midway Atoll, the young birds sing and dance with one another, gradually narrowing the field of possible mates until they finally settle on one. In the end, this "long engagement" usually takes several years, during which time the bond is strengthened as the pair "keep company" with each other: "dancing, and sometimes building nests, but not breeding" (Rice and Kenyon 1962:524).[4]

This elaborate courting process is one of the key mechanisms through which breeding is delayed in albatrosses. Thus extended courtship plays an important dual role in albatross breeding: by ensuring that young birds are a bit older and more capable before they breed for the first time, and by helping to produce partnerships that are solid enough to last the duration (Lindsey 2008:82).

Even after all this preparation, however, many of the chicks hatched each year do not successfully fledge. Many die as a result of dehydration and starvation (Sileo, Sievert, and Samuel 1990). In addition, threats such as storm surges and sand storms may cover or drown chicks and eggs. In an extreme case, such as the tsunami that followed the earthquake that struck Fukushima, Japan, in March 2011, tens of thousands of young birds can be lost in this way (BBC 2011). And then, finally, just as young birds are ready to take to the air, tiger sharks gather off the coast of each major colony in the area, ready to grab those who land or are blown into the water on their first tentative flights. While raising albatrosses of any species is no small feat, it is perhaps the Black-foots of Midway (who tend to nest in vulnerable sandy locations) that have it hardest of all, with no other species thought to face the same levels of predation and mortality in the first months of life (Lindsey 2008:98).

FLIGHT WAYS

It is through these arduous processes, full of obstacles and often fraught with danger, that albatross species have persisted through the vastness of evolutionary time. While we do not know precisely how long they have been riding ocean winds, it is certainly a considerable period. Recognizable petrels and albatrosses can be found in the fossil record from roughly 32 million years ago. As early as 9 million years ago, fossils from the Southern Ocean indicate that these birds were already very similar to their current forms (Jones 2008:143). Millions of years before anything like the human species appeared on the scene, albatrosses were already soaring, dancing, and fishing across this great blue planet.

Approached with an attentiveness to evolutionary history and a focus on the complex and difficult emergence of each new generation, it is clear that this thing we call a "species" is an incredible achievement. Each of the literally millions of generations of albatrosses that have followed one after the other has itself been ushered into the world through this narrow passage: laid, incubated, hatched, guarded and fed by parents, before taking those first steps toward flight and the world beyond. We often do not appreciate—and perhaps we cannot truly grasp—the immensity of this intergenerational work: the skill, commitment, cooperation, and hard work, alongside serendipity, that are required in each generation to carry the species through.

It is with recognition of this embodied intergenerational achievement that I understand species as "flight ways." This understanding is possible only since Darwin. Central to the mode of thought that evolutionary theory opened up is the transition from an understanding of a species as a fixed and eternal "kind" to that of something more akin to a historical lineage stretched between a beginning (speciation) and an inevitable end (extinction).[5] In this context, a species must be understood as something like a "line of movement" though evolutionary time. But it is much more than an empty trajectory. Each species lineage embodies a particular *way of life,* a particular set of morphological and behavioral characteristics that are passed between generations.[6] But this is also not a static way of life. More than the sum of those individuals currently living, species are engaged in an ongoing intergenerational process of *becoming*—of adaptation and transformation—in which individual organisms are not so much "members" of a class or a kind, but "participants" in an ongoing and evolving way of life.

In this context, any individual bird is a single knot in an emergent lineage: a vital point of connection between generations—generations that do not just happen, but *must be achieved.* The months and years spent cementing pair-bonds; the countless trips and thousands of miles flown by parents to provide for their chicks; the huge quantities of fish eggs, squid, and other foods that must be collected and carried back—*this* is the work that knots one generation to the next, that constitutes and preserves a species. What is tied together is not "the past" and "the future" as abstract temporal horizons, but real embodied generations—ancestors and

Laysan Albatrosses nesting at Midway Atoll. (David Patte/U.S. Fish and Wildlife Service; CC BY 2.0)

descendants—in rich but imperfect relationships of inheritance, nourishment, and care. These are knots *of* and *in* time—what Deborah Bird Rose (2012b) has called "knots of embodied time" (128–31)—connecting each generation to the next, and beyond to the broader emergent flight way.

WASTED GENERATIONS

But in our time, Laysan and Black-footed Albatrosses, along with so many other species, may be coming to an end. All this flight—all this dancing, incubating, and brooding to nurture all these generations—may be halted at the point where their paths cross our own ever-expanding, ever-more-widely circulating, impacts and demands on life systems.[7]

Today, the most significant threat to an albatross in the North Pacific Ocean (or any other ocean, for that matter) comes in the form of a fishing line and a baited hook. Bycatch estimates for Black-footed and Laysan Albatrosses vary, but annual mortality has likely been between 6,000 and 10,000 birds of each species for roughly the past fifty years (Arata, Sievert, and Naughton 2009:14–15; Naughton, Romano, and Zimmerman 2007:10). Since these birds ordinarily have low adult mortality and delayed reproductive maturity, these kinds of losses have a huge impact on the survival chances of these species (Ludwig et al. 1998:229). More than any other current threat, it is the persistent and widespread impact of fisheries on albatrosses that has caused them to become the most endangered family of birds on the planet (De Roy, Jones, and Fitter 2008:149).[8]

But, as we have seen, fishing is by no means the only obstacle that Midway's albatrosses currently face. Even here, in one of the most remote breeding places on Earth, these birds have been painfully exposed to the circulating detritus of many contemporary human societies. In particular, they encounter the plastic mentioned at the outset of this chapter, as well as a range of persistent pollutants like DDT (dichlorodiphenyltrichloroethane) and PCBs (polychlorinated biphenyls). These harmful materials are produced as chemical waste from industrial and agricultural processes and as plastic waste from the disposable and consumptive lifestyles of so many people around the Pacific. This is waste that implicates many of us—some much more than others—in the harm that it does as it builds up and circulates through the atmosphere, through rivers and out to oceans, and ultimately through the bodies of living beings.

In recent years, the accumulation of plastic in the world's oceans has become an increasingly topical issue. In the heart of the foraging range of Laysan and Black-footed Albatrosses lies perhaps the most publicized concentration: the so-called North Pacific Garbage Patch. "Patch" is not an entirely accurate term for this gathering. It is not a tightly packed and readily visible collection of junk, as the name perhaps implies, nor is it limited in size, in the way that one might imagine a mere "patch." Rather, it is a vast and shifting expanse of ocean in which the average concentration of rubbish is thought to be significantly higher than normal—so large, indeed, that one alternative name for the area does not seem wholly inappropriate: The Seventh Continent (Scarponi 2012).

Within this stretch of the ocean, plastic and other debris collects in various densities, shaped not only by the movements of large currents, but also by smaller-scale oceanographic features. One particularly important area for marine plastic is the Subtropical Convergence Zone (Kubota 1994; Pichel et al. 2007). Here, just north of the Hawaiian Islands, plastic items likely disposed of in countries all around the Pacific come together (Donohue and Foley 2007).

With so much garbage floating near the water's surface, it is not at all surprising that pieces of plastic and other debris find their way into albatross bellies. This situation has been vividly captured by the American photographer Chris Jordan in his series of photos "Midway: Message from the Gyre" (2009). Picture after picture depicts the decomposed bodies of albatross chicks—just bones, feathers, and a beak remaining, and in the middle of each, a multicolored pile of plastic and other debris: cigarette lighters, bottle tops, toy soldiers, and so many other little items. Seeing these photos, it is hard to imagine that the cause of death of these birds could be at all uncertain. It seems, however, that very few of them die as a direct result of consuming plastics (for example, through perforation of the digestive tract). Instead, ingested plastics likely contribute to other significant and common causes of mortality, such as dehydration and starvation (as chicks carrying heavy "plastic loads" are unable to eat healthy quantities of food) (Auman, Ludwig, Giesy, et al. 1997; see also Safina 2007). For those birds that do survive to fledge, suppressed appetites may well lead to reduced growth, and so to a lower fledging weight and decreased chances of long-term survival (Naughton, Romano, and Zimmerman 2007:14). Unfortunately, the further studies needed to establish this kind of indirect mortality are hampered by the difficulty of finding plastic-free birds to compare respective health and survival rates (De Roy 2008). As the sheer volume of plastic, both in the environment and in albatross bodies, continues to rise steadily, this situation can only become worse (Auman, Ludwig, Giesy, et al. 1997:243).

But before albatross chicks have even made it out of the egg, they are also exposed to a range of persistent organochlorines like the PCBs and DDT. These compounds have wide-ranging impacts on avian bodies, but many of their effects target reproductive processes. PCBs are known to reduce fecundity in birds through embryo mortality and to interfere with neurological development, endocrine function, and cell growth. DDT and its metabolites cause eggshell thinning (Auman, Ludwig, Summer, et al. 1997), which can ultimately lead to the loss of eggs that are accidentally crushed by incubating parents, as occurred most famously in the United States with the Bald Eagle (*Haliaeetus leucocephalus*).

Making their way through rivers and streams to the ocean, or traveling in the atmosphere, these toxic residues of our industrial societies circulate endlessly through the environment, accumulating in those unfortunate places where particular constellations of temperature, wind direction, water current, and landform deposit them. Although their use has for decades been banned in many parts of the world, they continue to linger, in some cases metabolizing into equally toxic compounds (like DDT into DDE [dichlorodiphenylethylene]). Along their journey, and particularly in the places where they become concentrated in the environment, they are ingested and absorbed, building up in the fatty tissue of living organisms. From here, they inevitably move up trophic levels within food chains, so that at each level concentrations multiply (in a process known as "biomagnification"). As a result, top-level predators like albatrosses end up carrying concentrations of these compounds that are orders of magnitude higher than those found in the surrounding environment.[9]

Several studies indicate that Black-footed Albatrosses have PCB and DDT levels that are known to cause eggshell thinning and embryo mortality in other fish-eating birds (Auman, Ludwig, Summer, et al. 1997; Guruge, Tanaka, and Tanabe 2001). Myra Finkelstein and her colleagues (2007) have also noted that these levels are "of comparable magnitude to concentrations associated with reproductive deformities and impaired immune function in birds from the Great Lakes" (1896), although we do not as yet know whether similar levels will produce similar physiological effects. But eggshells do appear to be thinning: eggs collected in 1995 were 34 percent thinner than those collected before World War II (Ludwig et al. 1998). While it is very difficult in toxicological studies like these to measure physiological changes in a wild population as the result of a single contaminant (Finkelstein et al. 2007:1897), it is increasingly clear that Black-footed Albatrosses are in the thick of a perilous and toxic space. Like that resulting from the accumulation of plastics in the North Pacific, this situation is made additionally worrisome by the fact that contamination levels are increasing; the toxic burden in both Black-footed and Laysan Albatrosses is much higher than it was even a decade ago (Finkelstein et al. 2006).[10]

And so, in forms both visible and invisible to the naked eye, the waste of human societies circulates in vast atmospheric systems and oceanic currents to accumulate in albatross bodies. Chemicals, plastics, living beings and their lineages, commitments, relationships, and much more are brought together here in a tangled web of interactions. Temporalities converge in this meeting of bodies, each carrying histories and presaging futures inscribed in them by evolutionary inheritances and/or the processes of their design and manufacture. Millions of years of albatross evolution—woven together by the lives and reproductive labors of countless individual birds—comes into contact with less than 100 years of human "ingenuity" in the form of plastics and organochlorines discovered or commercialized in the early decades of the twentieth century. (Of course, these products are themselves grounded in yet other histories—for example, those of the ancient fossilized remains that over millions of years became the oil used to produce plastics.) In their current forms, these human-engineered products linger and accumulate to play their part in the undoing of the intergenerational achievement that is the albatross flight way. Perversely, and as if to highlight the fragility of the intergenerational bonds that constitute species, the impact of these toxic substances is felt almost entirely by breeding birds and their young; it is felt at that precise point where one generation brings forth the next. As shells break and young birds die, the next generation of albatrosses fails to fledge, and the knots that should bind the past to the future life of the species come undone.

But while these toxic products have a short *past* (at least in their current forms), their *future* is not limited in the same way. Rather, there is something almost immortal about them. In Timothy Morton's apt terms, they are "hyperobjects": "objects that are massively distributed in time and space relative to humans" (2012:81), "products such as Styrofoam and plutonium that exist on almost unthinkable timescales" (2010:19). We just do not fully understand how long they will endure and what forms they may take as they move through time. While plastics "break down," they do not disappear. Instead, they become "micro-plastics" whose smaller size allows them to enter and accumulate in smaller and smaller bodies, so they might gradually affect a larger and larger range of living beings (Barnes et al. 2009). In fact, with the exception of those plastics that have been incinerated—to contribute to other toxic legacies—all those ever produced are still around in one form or another (Barnes et al. 2009), ensuring that countless future generations of albatrosses, humans, and other organisms inherit a growing problem.

Temporalities get messy here. Within the huge time frames needed to think the evolution of species and the lives of plastics, the daily struggles of individual birds disappear from sight. The work that they do to hold together generations does not even register—until it abruptly ends. From this temporal perspective, what is most startling and disturbing about the current impacts of plastics and organochlorines on Midway's albatrosses is the terrifyingly fast pace at which they are occurring. But if we "zoom in," to view the world from the perspective of a finite living being, the current deaths of albatrosses instead become torturously drawn out, a perfect example of the kind of "slow violence" that Rob Nixon (2011) has described: "a violence that occurs gradually and out of sight . . . dispersed across time and space" (2). This is a violence that is often rendered invisible precisely by its slowness, by the *gradual* accumulation of toxins in the environment and in the tissues of fleshy bodies that eventually pushes things beyond the point where life is possible. In contrast to the bloody immediacy of gunshots and bomb blasts, which are readily recognized for the violent actions that they are, Nixon points out that the continued use and abuse of substances that *we know* will gradually accumulate to kill others flies under the radar to such an extent that we often cannot even identify it as a form of violence at all.

In exploring the plight of albatrosses in our time, we are drawn into an encounter that is situated across multiple temporal horizons and scales. In this context, plastics and synthetic chemicals emerge as both recent and enduring forces, impacting the world around them in a way that is simultaneously frighteningly rapid and painfully slow. Similarly, species appear both as vast evolutionary lineages and as a collection of fleeting and fragile individual birds, doing the mundane work of knotting together generations. In some sense, millions of years of evolution are "in" each of these albatross bodies: inheritances, histories, relationships, carried in the flesh.

Encountering the world through an attentiveness to divergent and overlapping temporal frames is not always easy or straightforward, but doing so enriches our understanding of these troubling times. Inspired by Michelle Bastian's (2011, 2012, 2013) work on time and community, I am interested in the way in which paying attention to the diverse temporalities enfolded in the encounter between an albatross and a sea of plastic invites us into different relationships, different understandings, and ultimately different responsibilities. This is not about making time neat, flat, and singular. Rather, consciously inhabiting multiple, conflicting, and intersecting temporalities brings its own challenges and possibilities. And so I am interested in how rethinking albatrosses as beings that emerge from and live and die within dense webs of overlapping temporalities and inheritances remakes our understanding of the immensity of what is lost in extinction, while drawing us into new and deeper responsibilities.

FLIGHT WAYS TO EXTINCTION

Sitting among a tiny colony of Laysan Albatrosses on the north shore of the island of Kaua'i in December 2011, I was again reminded of the deep historical roots of this form of life.[11] The breeding birds in the colony at that time were probably mostly males, taking their first long shift on the egg as their partners headed out to sea to feed and recuperate. For the most part, these incubating birds sat quietly in a seemingly trance-like state. And yet the colony was a lively place. All around these older birds, juvenile albatrosses were busily dancing and singing with one another,

A Laysan Albatross with its new egg on the island of Kaua‘i. (Photograph by author)

cementing the relationships that may one day become pair-bonds. On land and in the air, these young birds playfully interacted with one another through movement and sound.

As we spent time among the colony, sometimes sitting or standing within a few feet of nesting birds, we were occasionally greeted by the cautious clacking of a nearby bird's beak. After a few seconds, though, the bird invariably turned away, seemingly uninterested in our presence. There is something slightly disconcerting about being in close proximity to a wild animal that has little interest in you.[12] As a human being, as a member of a species that is often watched very closely by others for potential danger, the seeming trust of albatrosses felt somewhat odd. Of course, "trust" is not the right word here. Rather, the situation is one in which human beings—like terrestrial mammals in general—have not been a relevant environmental factor for most of the long evolutionary history of this species (or of so many other colonial birds that have historically nested on islands free from predators).[13]

Standing in the presence of albatrosses, we are required to occupy a strange position. Through their behavior, we are reminded of the *long* duration during which we as a species mattered so little to the fates of others that we simply did not—and still do not—register as a relevant feature of their world. And yet, we must also stand in their presence in the full knowledge that at the present time nothing could be further from the truth: "we" are now the single greatest threat—and on multiple fronts—to the possibility of the continuity of albatross generations. Standing in their presence, we are required to somehow inhabit both this long past and this tragic present. In short, we are drawn into an awareness of the immensity of our "geological moment."[14]

It is in this context that approaching species as "flight ways" may "enable a recognition of [the] multiple, contradictory allegiances within the 'same' moment" (Bastian 2011:165). In short, it may allow us to hold the fleeting lives of individual birds in tension with the life of their larger species, to inhabit a time of extinctions in full consciousness of the vast periods that have been required to produce those living forms now being lost.

Doing so enables us to move beyond the simplistic assertion that we need not worry about the extinction of the Laysan and Black-footed Albatrosses because extinction is an unavoidable reality in the life of all species. Just as evolutionary theory enabled new understandings of species as historical lineages, post-Darwin our understanding of the process of extinction has also taken on a new light. Evolutionary theory introduces us to a world of constant adaptation and change in which extinction, far from being an aberration in a fixed and timeless order, is an unavoidable part of the life cycle of all species. In this context, while albatrosses may pass out of the world sooner than they would have without human involvement, their extinction was always an inevitability.

The principal problem with this understanding is that it is grounded in a singular evolutionary time frame that provides little insight: From the perspective of eternity—or "Life," whatever that might be—what does an individual organism, or a species, really matter?[15] In short, this is a perspective that covers over the many other possibilities, relationships, and responsibilities coalescing in any given moment, possibilities that become visible only when we also practice an attentiveness to other temporal frames.[16]

This attentiveness is an effort to "hear anew" the Darwinian revolution, as James Hatley has suggested we must in the current period of mass extinction.[17] It is an effort to find new ways to inherit, inhabit, and bequeath the worlds that evolution describes. To this end, this chapter has focused on the work of reproduction and the transition between albatross generations: the capacity of these birds to take flight and move tirelessly across vast expanses of open water in search of food, and to lay eggs, brood chicks, and fledge young for months at a time, year in and year out. In so doing, I have highlighted the fact that species do not just happen, but must be *achieved* in each new generation, held in the world through the labor, skill, and determination of individual organisms in real relationships of procreation, nourishment, and care.

Of course, many other species do not rely on such incredible individual efforts to maintain their continuity across generations. Fish spawn, passing eggs and sperm into the water en masse, never to be encountered again (although some—Pacific salmon, for example—must undertake long and difficult journeys upriver and sacrifice their own lives before doing so). Flowering plants are often pollinated by wind, birds, or insects and then produce seeds that simply fall to the ground or are carried away by other animals. In the reproductive efforts of these and many other species, there is not the same readily recognizable "work" that we encounter in the life cycles of albatrosses. In many other species, these reproductive efforts are also not divided into heterosexual roles (or sexual roles at all, for that matter); even among North Pacific albatrosses, there is a great deal of diversity in the ways in which reproductive roles are taken up.[18] As in all other facets of evolution, when it comes to reproduction, life evinces an incredible diversity of approaches and strategies. And yet, in each case a specific form of life is produced through a long evolutionary history of adaptation in which individuals invest time and/or precious resources in a generational continuity that stretches out beyond them. In the albatross, we find an extreme—or, perhaps, just a relatively familiar—manifestation of a drive common to all speciated life forms.[19]

Individual participants in a species may not be aware of where they have come from or where they are going in this broader evolutionary sense, and yet, it is *their* striving for continuity that achieves it nonetheless. This striving is, of course, present in the individual organism's efforts to maintain itself, but for our purposes what is equally important is the efforts of living things to pass on that life to subsequent generations (Margulis and Sagan 1995:213; Rose 2006). Each year, taking on the dangerous and demanding obstacles of reproduction, albatrosses, like so many of Earth's other creatures, invest their lives in a generational continuity that exceeds them.

In the language of ethics, this striving expresses a clear "interest" on the part of the individual for not only its own continuity, but that of subsequent generations. But a living being and its offspring cannot survive and thrive on their own. For speciated forms of life, like albatrosses, intergenerational continuity requires at a minimum the larger reproductive community that is a species. And so genuine respect for the striving of individuals must necessarily be coupled with a respect for the broader species.[20] This respect is the basis of an ethic that does not attempt to draw a line between those beings worthy of moral considerability and those outside such a requirement. Rather, it is the basis of a practice that respects all life, a practice in which, in the words of Shagbark Hickory, "considerability goes 'all the way down'" (quoted in Plumwood 2011:206).[21]

Bound together by shared inheritances and a shared fate, a species is always more than its current manifestation. With this in mind, extinction is the loss not of a single fixed "kind," but of a potentially limitless set of emergent and branching flight ways from the present into the diversity of the future. Each species is ultimately a flight way *beyond itself*. Through ongoing patterns of speciation and phyletic evolution, a species is always becoming different from, other than, itself. And so what is lost in extinction is not "just" the current manifestation of a flight way—a fixed population of organisms—but all that this species has been, as well as all that its past and present might have enabled it to one day become.

These are the flight ways into the deep past and future that are contained within the emergent becoming of a species. Generations stretching before and after, beyond the horizon of our temporal comprehension. It is all their work and their possibilities that are at stake in the living and dying albatross bodies that populate this blue planet at any given "now." And so it is under the shadow of *all* these generations, of a flight way taken—however impossibly—in its entirety, that we must feel the weight of an ethical claim to hold open space for the continuity of this ancient and evolving form of life.

ENTANGLED FLIGHT WAYS

The albatrosses of Midway are not the only species for whom extinction now looms ever larger. Even in their immediate vicinity, numerous other species are positioned perilously in that space at the edge of life (Steadman 2006; Stearns and Stearns 1999). Today, we find ourselves deep within a period of mass extinction. This is a time of loss in which a respect for the continuity of ways of life takes on both a new intensity and a heightened sense of urgency. As discussed in the introduction to this book, with current rates of extinction estimated to be somewhere between 100 and 1,000 times "normal" background levels, the current loss of Earth's diversity may be on a scale not experienced since the extinction of the dinosaurs roughly 65 million years ago (Aitken 1998; Primack 1993). In short, we may be ushering in the sixth mass extinction event since "complex" life evolved on this planet (and, with it, the fossil record that enables paleontologists to sift through life's history and identify extinction events).

In the context of this period of incredible loss, the cumulative deaths of so many species take on a new kind of destructive possibility, threatening to overrun life as we know it and usher in a new time. While the diversity of life on Earth has rebounded after each of the five previous mass extinction events, it has done so each time in a drastically altered form—just as the extinction of the dinosaurs made room for the mammalian radiation that produced so much of the diversity of life found on Earth today, including ourselves (Mayr 2001:133). And so, while mass extinction events do not undermine "Life" as an enterprise in any objective sense—over time spans of millions of years, living systems will (likely?) recover and diversify—from the perspective of any single Earthling bound within a *particular* ecological and evolutionary community, things look very different. As a living being, a species participant viewing the world from the inside, mass extinction is the radical transformation to the point of death of whole complex life systems. When the dust settles, something will likely recover, but it will be something very different.

Situated *inside* this time of extinctions, the particular nature of the ethical claim made on us to hold open space for other species requires an understanding of the complex histories and inheritances that draw us into responsibility and relationship with others. A central part of this history is our own evolution as a particular social mammal: a species with the intellectual and affective *capacity* to care for others both of and beyond our own species, to recognize their interests and act in ways that make room for our combined flourishing.[22] In paying attention to evolution, we are reminded that this capacity is one with a pre-human history. Empathy, consideration, care, even various form of ethics have never been the privileged possessions of humanity, and they certainly did not arise as part of the civilizing emergence of cultured humans radically distinct from a nature that is "red in tooth and claw." Rather, these kinds of affective engagements—these ways of being meaningfully with others—are products of a long and complex inheritance that undermines any simple distinction between the "biological" and the "social."[23]

Acknowledging that human values and experiences are grounded in specific biologies and evolutionary histories does not necessarily lead to crude biological determinism of one sort or another. As Donna Haraway (2004) notes, "Biological-determinist ideology is only one position opened up in scientific culture for arguing the meaning of human animality. There is much room for radical political people to contest for the meanings of the breached boundary" (10). In short, we must stop being scared of our biology and find new ways to think about our constitutive entanglement in multispecies worlds; new ways to figure "human nature" as something that, as Anna Tsing (2012) notes, has "shifted historically together with varied webs of interspecies dependence." In short, new ways to acknowledge that "human nature is [and will always be] an interspecies relationship" (144).

In this context, our evolution as a species is not the only inheritance that lays us open to the ethical claims of others. We are also heirs to particular cultural and economic histories and practices: projects of development, modernization and profit maximization. Intricately entangled with these are the social and technological processes that have yielded diverse toxic chemicals and a world of brightly colored, disposable but not fully degradable, plastics. As a result, in recent decades we have also become heirs to Rachel Carson's legacy and the emergence of the modern environmental movement. We have become members of a society with the capacity to recognize the specific harms that our ways of life are producing and to reflect critically on the consequent ways in which we are undermining the sustainability of our living planet. As Timothy Morton (2011) succinctly put it: "This is the historical moment in which hyperobjects become visible by humans. This visibility changes everything."

And so, as a result of diverse, interwoven patterns of inheritance over immense temporal horizons, we have emerged as beings whose lives are lived "in the shadow of all this death" (Rose 2012a), but simultaneously under the weight of an ethical claim to work toward more "livable worlds" (Haraway 1997). In this context, it is our very capacity to recognize the interests of others and alter our behaviors accordingly that makes a claim on us, that makes us the kinds of beings whose lives *must* be lived within the space-time of such a claim. We are, of course, free to negotiate the nature of this claim and the obligations that it produces, but we must do so in a way that acknowledges the particular histories and entanglements that constitute our specific forms of life.

In this context, an ethical response to this period of massive loss requires more than respect for any individual species. In addition, it is about the survival of myriad evolutionary lineages that together form our entangled multispecies community. In short, individual flight ways must be understood within their broader contexts: within what Deborah Bird Rose (2012b) has called "knots of embodied time" (128–31). Rose argues for the need to recognize patterns of both "sequence" and "synchrony" in Earthly life's movement through time. The flight ways that have been the subject of this chapter thus far are patterns of sequential, generational, life.

But synchrony asks us to be attentive to the way in which the multiple and diverse flight ways that constitute Earth's diversity are also delicately interwoven *with one another*. The Black-footed Albatross, like any other species, is not a flight way through an empty void, but an entangled way of life, bound up in and becoming as part of a specific multispecies community. In Rose's (2012b) terms: sequence "involves flows from one generation to the next. Synchrony intersects with sequential time, and involves flows amongst individuals, often members of different species, as they seek to sustain their individual lives" (129).[24]

And so there is an important sense in which, in addition to being carried through time by the efforts of their own generations, species also carry *one another*, nourishing and being co-shaped as members of a particular entangled community of life. If we abandon the seeming objectivity and impartiality of a "view from nowhere" (Haraway 1991), and affirm our place as mammals, as beings "ecologically embodied" (Plumwood 2003) and woven into patterns of mutual sustenance and co-evolution with other species, then the current period of mass extinction can be regarded only as an assault on life. While the future of our own species is surely at stake, our response to the ethical claims of mass extinction cannot be understood through the lens of a simplistic anthropocentrism. Rather, it must be premised on something more akin to a "Cenocentrism"—a plea for the continuity of the "Cenozoic achievement," for the continuity of a whole evolutionary and ecological community that reaches back into the distant origins of life on Earth, but that takes its primary form from the specific period after the last mass extinction event: Cretaceous–Tertiary (K–T) event.[25]

Through the clouds of dust and smoke that likely covered much of Earth during this time, several species of birds made their way into a new world. Scientific debate continues over precisely how many avian species survived the K–T event, with estimates varying widely (Cooper and Penny 1997). We know, however, that some birds made it through this dark time. Alongside them, some mammals, reptiles, plants, bacteria, and others also survived, and it is from these individuals that the incredible diversity of present-day life on the planet evolved. This is the community of life that produced our own species, the community to which we belong. It is not a static, atemporal ecology, but an infinitely complex set of entangled flight ways—countless generations of all kinds holding themselves and each other in the world.

And so we are now ourselves placed under the weight of a collective ethical claim made on us by *all* these generations, by all the living things that have populated this planet over the past many millions of years, as well as all those that might yet come—in "endless forms most beautiful and most wonderful" (Darwin [1859] 1959). Thus perhaps we are called to account by nothing less than the entirety of life on this planet, for all the ways in which, during our own brief lives, we help to shelter or destroy the entangled diversity of forms through which life makes itself at home in our world.

For this briefest of evolutionary moments, *Homo sapiens* has come into direct and profound contact with the world's albatrosses and countless other species. Our flight ways have crossed, and each of us has become implicated in the fate of the other. Standing among a colony of albatrosses along a windy cliff top on the island of Kaua'i, I was reminded of the shifting and consequential nature of this entanglement. As nesting birds quietly ignored my close presence, I reflected that perhaps what is most tragic about the current situation is not the "failure" of albatrosses to adjust or adapt to new threats and an altered environment: intensive long-line fishing or brightly colored plastics that look like food. Rather, what is most tragic is another failure to adapt. Our own failure—in which some societies and some people are far more complicit than others—to come to terms with our own relatively new capacity to systematically alter environments in a way that undermines possibilities of life for other living beings and, ultimately, for ourselves. Perhaps it is we who have not yet "evolved" into the kinds of beings worthy of our own inheritances.

Two

CIRCLING VULTURES

Life and Death at the Dull Edge of Extinction

The long-billed vulture, *Gyps indicus*, is one of three vulture species that serve as sanitation engineers in India, Nepal and Pakistan. For thousands of years, they have fed on livestock carcasses. As many as 40 million of the birds once inhabited the region. Obstreperous flocks of vultures thronged carcass dumps, nested on every tall tree and cliff ledge, and circled high overhead, seemingly omnipresent.

SUSAN MCGRATH, "THE VANISHING"

In conversations about vultures in India, people have often recounted to me having seen large numbers of these birds gathered along the banks of rivers, consuming the dead bodies of cattle and other animals, including sometimes people, as they float by or wash up on the water's edge. When it meets a vulture's beak, it matters very little if this flesh, this meat, was once a human or some other kind of animal. In fact, numerous human societies throughout history—including current-day Parsee communities in India and Buddhists in Tibet and elsewhere—utilize exposure to vultures as the most appropriate way of "taking care" of their dead (Schuz and Konig 1983; Subramanian 2008; van Dooren 2011b).

I am interested in the dynamics and practicalities of eating and being eaten in multispecies communities. Eating is, of course, one of the most important ways in which the dead are woven into the lives of the living. But there are many other important entanglements, too, many other ways in which the dead—through the active presence of their decaying bodies or the simple absence of their living participation—help to shape the world in which we all, for better or worse, make our lives. This chapter explores one such entangled community of the living and the dead—a

A Long-billed Vulture perching in Orchha, Madhya Pradesh. (Yann; CC BY-SA 2.0)

community that centers on critically endangered Indian vultures, but draws in humans, cattle, and a host of others. I am interested in how these multispecies entanglements were formed and now, as vultures disappear from India, how they are being unmade with disastrous consequences for everyone. In taking up these issues, this chapter presents an understanding of extinction as a drawn-out and ongoing *process* of loss: the "dull edge of extinction." Even in the case of India's vultures—who have shocked many with the speed of their population decline—extinction must be understood as the slow unraveling of entangled flight ways.[1] In contrast to conventional ways of thinking about extinction—according to which its precise moment of occurrence is pinpointed in the death of the "last individual"—India's vultures remind us that extinction is often characterized by enduring patterns of suffering and death that begin well before the death of the "last vulture" and will likely continue long afterward, haunting future possibilities for a host of living beings.

Vultures are our guides into this space of loss, death, and extinction. These curious creatures are relatively unique among the world's birds and mammals in that they are "obligate" scavengers. This means that unlike most other scavengers, vultures do not opportunistically alternate between predation and scavenging, but rely (relatively) exclusively on finding animal carcasses. A vulture's whole body is oriented and adapted to this method of food procurement and the lifestyle that accompanies it. Indeed, biologists Graeme Ruxton and David Houston (2004) have even argued that it may have been in the development of the large body size and specializations for efficient soaring flight that are so essential to being a successful avian scavenger that vultures like those found in India lost the flying accuracy, agility, and maneuverability that are still found among the other "birds of prey."

However it evolved, though, scavenging has been a highly successful way of life for vultures for a very long time. While the fossil record is poor, it seems that in the "Old World," vultures stretch back more than 20 million years. The genus *Gyps,* to which the vast majority of the vultures in India belong, probably arrived on the scene sometime in the last few million years (Houston, pers. comm.; Rich 1983).[2] But while scavenging has been evolutionarily successful for vultures, it is arguably not the most attractive way of getting a meal. It seems fair to say that vultures are a little bit gross. The *Gyps* species that occur in India live primarily in colonies, usually of 20 to 30 birds but sometimes in excess of 100, often roosting as close as possible to dumps or slaughterhouses, building their nests in tall trees or on cliff ledges, and lining them with wool, skin, dung, and rubbish (Ferguson-Lees and Christie 2001:422–28). In addition, the focus and vigor with which vultures have taken to scavenging has meant that they often encounter less than "fresh" food and, as such, are required to possess a high level of resistance to various pathogens and diseases (Houston and Cooper 1975).[3] According to Dean Amadon (1983), "It has been reported that vultures can, without ill effect, consume the viruses of such deadly pathogens as anthrax in such quantities as would fell an ox—or a whole heard of oxen" (ix).

But something is now poisoning vultures and threatening to cause their extinction in India and throughout the surrounding region. Over the past two decades, vultures have been dying en masse, largely as a result of their being unintentionally poisoned by a drug called diclofenac, which is given to cattle whose carcasses vultures eat. In vulture bodies, diclofenac causes painful swelling, inflammation, and eventually kidney failure and death. Today, it is thought that approximately 97 percent of the three main species of vulture in India are gone (Prakash et al. 2007; Swan, Cuthbert, et al. 2006). The discovery that vultures were disappearing entered the scientific literature through the work of Vibhu Prakash (1999) of the Bombay Natural History Society (BNHS). Since then, the BNHS, in collaboration with the Royal Society for the Protection of Birds in the United Kingdom and the Zoological Society of London, has been conducting further research on the decline of vultures in India. Together, these organizations have also established several conservation and breeding centers in India. The hope is that one day, when the threat is gone and there is a large enough captive population for it to be sustainable in the wild, the vultures might be able to be (re)released. Whether this will ever happen, however, remains an open question.

BETWEEN DEATH AND LIFE (THERE IS SOMETIMES A VULTURE)

It is a strange thing for an animal that is so closely associated with death, albeit usually the deaths of others, to be itself on the way to extinction.

Like many other people throughout history, and around the world, it seems to me that the vulture is a kind of liminal creature, inhabiting a space somehow strangely between life and death. Perhaps it is the way in which vultures seem to sense death, often arriving before it occurs (at least in part a skill acquired through good vision, the ability to search far and wide, and a kind of communicative observation of other vultures [Jackson, Ruxton, and Houston 2008]), or maybe it is because they consume the dead and then soar high up into the sky (Houston 2001:51). The deathly association and liminality that interests me, however, is the vulture's ability to "twist" death back into life (Rose 2006). My thinking here is situated within the kind of ecological context that Deborah Bird Rose and Val Plumwood have outlined. They remind us that death must be thought about not as a simple ending, but as completely central to the ongoing life of multispecies communities, in which we are all ultimately food for one another (Plumwood 2008b). As Heraclitus succinctly put it: "the one living the other's death, and dying the other's life" (quoted in Plumwood 2011).

In this context, vultures are at the heart of life and death's transformative potential. But instead of taking life to produce their nourishment, they consume only that which is already dead, pulling dead flesh back into processes of nourishment and growth. I suspect that alongside the insects, bacteria, fungi, and other organisms that also make their living breaking down the dead, vultures have a special place in life's heart. I cannot help but think here about Jean-Luc Nancy's (2002) beautiful injunction not to separate life from death: "To isolate death from life—not leaving each one intimately woven into the other, with each one intruding upon the other's core [*coeur*]—this is what one must never do" (5). Vultures understand this intimate entanglement of life and death. I think that they would appreciate Nancy's sentiment and feel a deep bodily agreement with Georges Bataille (1997) when he noted that "life is a product of putrefaction, and it depends on both death and the dungheap" (242).

Death, understood in this way, positions all organisms (including humans, a point that should not have to be made, but unfortunately often does) as parts of a broader multispecies community. Possibilities for life and death, for everyone, get worked out inside these entangled processes of "becoming-with" (Haraway 2008). That we can live at all, but also that we live in the *ways* in which we do, is the result of our specific situatedness in a more-than-human world. Understandings grounded in "human exceptionalism" will be of no help to us here (Plumwood 2007). The biological and the social, the material and the discursive, alongside the living and the dead, are all mixed up in the formation of what come to be (and to count as) bodies, societies, religions, cultures, and ecologies. The interactions of people, vultures, and others in India highlight some of these tangled processes of becoming, as well as the life and death stakes of the specific ways in which we are bound up with and exposed to others in a time of increasing environmental change.

ENTANGLED BECOMINGS

Three main species of vulture make their homes in India, and all three are of the genus *Gyps*: the Oriental White-backed Vulture (*G. bengalensis*), the Long-billed Vulture (*G. indicus*), and the Slender-billed Vulture (*G. tenuirostris*). In the first half of the twentieth century, these species could be found in large numbers right across Southeast Asia, throughout the Indian subcontinent, and into Pakistan. During the second half of the century, however, populations began to disappear throughout Southeast Asia, the eastern part of the range. The causes of this population decline are not known with certainty, but perhaps the most significant factor was the loss of reliable food sources. This loss likely occurred as a result of both the collapse of wild ungulate populations (through uncontrolled hunting) and changes in the husbandry of domestic animals. In addition, direct persecution of vultures by humans is thought to have had an impact on their numbers, as did poisoning and habitat loss in some local areas (Pain et al. 2003:661–62). While there are still some small remaining pockets of vultures in Cambodia and perhaps Laos and Vietnam, *Gyps* vultures are now otherwise considered to be extinct in Southeast Asia (Pain et al. 2003).

In the face of these local extinctions, India was one of the last strongholds of these species. Throughout most of the second half of the twentieth century, during the period of decline in Southeast Asia, life was good for vultures in India. In 1985, Indian populations were still so large that some speculated that the Oriental White-backed Vulture was "possibly the most abundant large bird of prey in the world" (Pain et al. 2003:661). In India, there was nothing like the food shortage for vultures that occurred to the east—quite the opposite. While vultures in India certainly benefited from its being one of the most cattle-rich countries in the world, from a vulture's perspective what makes India an ideal place to live is that most of the cows there are not consumed by local people. Hindu reverence for cattle, alongside a more general ethos of *ahimsa* (nonviolence toward living things), has produced a unique and complex environment in which most Indians do not eat beef and many are vegetarian (although Muslims and a growing number of Hindus do eat animals, including sheep, goats, and sometimes cows [Robbins 1998]).[4]

Cattle are used in India predominantly for plowing, milking, and as general beasts of burden, and their dung is widely used as fuel and fertilizer (Robbins 1998:226). When they die, cows are usually either taken to carcass dumps or left at the edge of villages, often after being skinned for leather (Singh 2003). By and large, however, it is vultures that have been relied on in India to "take care" of an estimated 5 to 10 million cow, camel, and buffalo carcasses each year: "As many as 100 vultures may feed on a single cow carcass, stripping it clean in 30 minutes. Two thousand, 3,000, even 10,000 vultures swarmed the larger dumps in the early 1990s, the huge birds lapping at carcasses with their leathery tongues, thrusting their narrow heads neck-deep to reach internal organs, tussling over choice gobbets of meat" (McGrath 2007). In this context, vultures often lived quite closely with humans. In urban and semi-urban environments, they found abundant food at carcass dumps, as well as tanneries, slaughterhouses, garbage dumps, and bone mills (where they could pick the bones clean before they were crushed for use in fertilizers). But it was not just vultures that benefited from this association. These industries, and local communities, were provided with a free and efficient means of carcass disposal for the millions of cows that they kept but did not eat (as well as the waste products from numerous other kinds of animals).

Consuming the dead is, of course, what vultures do. In taking up this role, they help to stem the spread of contamination and disease (such as anthrax, which is endemic in parts of India). When they live closely with people, especially in urban environments, they provide an incredibly valuable "service" to human communities, as we will see. Understandably, this symbiotic exchange has provided an ideal situation for people and vultures to live side by side in India. Writing in 1983, Robert B. Grubh noted,

"Where a regular food supply exists, it is a familiar sight to see 200 to 400 individuals sitting crowded on trees or on rooftops in the vicinity" (108).

This close entanglement of vultures and people in India is made all the more remarkable when contrasted with other places around the world where *Gyps* vultures have little association with human populations. In East Africa, for example, people understandably do not make meat available to vultures, but they also often shoot them if they start visiting towns and cities—especially the larger vulture species (Houston 2001, pers. comm.). In addition to food availability, this contrast with Africa points to another important difference in the direct treatment of vultures by people. In India, vultures are far less likely to face the persecution that has afflicted them in so many other parts of the world—not just Africa and Southeast Asia, but also Europe and the Americas (Ferguson-Lees and Christie 2001). It is most likely that significant cultural and religious dimensions of Hinduism and life in India account for at least part of this differential treatment—for example, the birds' association with the mythical Hindu vulture king, Jatayu (Baral et al. 2007:151). There is most definitely also a more pragmatic dimension to this treatment, with vultures being widely understood to provide a very valuable service through the removal of dead flesh (Agence France-Press 2007; Houston, pers. comm.).

But India is anything but a monolithic "cultural landscape." While it is beyond the scope of this chapter to consider the diverse range of relationships with vultures around the country, outside of the majority Hindu culture, India's small Parsee community (centered in Mumbai) requires consideration in any discussion of vultures and their people in that country. For hundreds of years, Parsees in India have laid out their dead to be consumed by vultures in *dakhmas* (towers of silence), believing that dead flesh pollutes fire, water, and air—all of which are sacred (Subramanian 2008; Williams 1997:158). In this context, people provided an additional source of food for vultures, while the vultures provided an efficient and hygienic means of disposing of the dead (van Dooren 2011b). This Parsee practice provides a wonderfully rich acknowledgment and reminder of the place of human flesh in multispecies nourishment. Of course, Parsees and vultures do not have to be told that humans are edible. Some members of the Parsee community were so committed to the role of vultures in taking care of the dead that when vulture numbers at towers of silence

A Long-billed Vulture in flight. (Vaibhavcho; CC BY-SA 3.0)

started to decline, they proposed setting up a small captive population in an aviary to continue the tradition.

I have offered only a handful of general examples from a diverse history of human–vulture interactions. What is clear, however, is that India has provided a somewhat unique environment for these birds. It is tempting to view this as a situation in which humans "accommodated" vultures, making room for them within their communities. In reality, however, generations of *Gyps* vultures stretch back several million years into the deep past of the Indian subcontinent. They were there well before human habitation, before humans even emerged as a species. It is likely that vultures evolved in close relationships with the wild migratory ungulate species like blackbuck, which were abundant until the arrival of people and domestic animals (Houston 1983, pers. comm.), only later settling into growing human societies, which, as Dominique Lestel, Florence Brunois, and Florence Gaunet (2006) note of "human societies" in general, "are precisely *never* exclusively human" (159). While it is certainly true that Indians have held open spaces for vultures in ways that other cultures around the world have not, it is also true that these people and their cultural and religious practices *emerged* and took shape in a land already inhabited by vultures.

Together, vultures, people, cattle, and others co-produced a unique environment in which food was made readily available for vultures, and, in turn, people were provided with a reliable and inexpensive means of disposing of the dead—this being particularly important for people who kept an abundance of cattle that they did not eat. In fact, one wonders how the cattle-keeping practices that have emerged in India in company with Hinduism would have looked, or perhaps even been possible, in the absence of these dedicated scavengers.

PROXIMITY AND "DOUBLE DEATH"

But now these vultures are dying. In an age of diclofenac, both cattle and human bodies often no longer provide nourishment; vultures are no longer able to twist death back into life. Instead, dead bodies are now poisoning vultures, producing more and more death. At the Parsee towers of silence, the decline of vultures seems to have begun in the 1970s, perhaps due to factors that include urban growth in Mumbai and an abundance of other food sources (Houston, pers. comm.), but perhaps also because humans started to take diclofenac in the 1960s, so the first birds to be affected would likely have been those that fed at the towers of silence. But, for vultures, the more recent widespread use of diclofenac for the treatment of cattle has been a far more serious problem. Diclofenac is used to treat any number of conditions in cattle, including lameness, mastitis, and difficult birthing (Cunningham, pers. comm.; Swan, Naidoo, et al. 2006:0395).[5] While the use of diclofenac in humans might have been catastrophic for a captive population of vultures at a tower of silence—one of the reasons why some scientists withdrew their support for keeping vultures in an aviary there—the use of the drug in cattle is far more significant for the overall collapse of vulture populations in India (because cattle are a far more important food source). As such, it is the vultures' role as consumers of cattle, rather than of people, that will be the primary focus of my discussion.

Importantly, the use of diclofenac in the treatment of cattle is often driven by poverty and the need to keep working animals even when they are old and sick. As Andrew Cunningham (pers. comm.) explained: "If an animal is going sick, is going downhill, they want to get the most out of the animal. . . . So you just pump pain killers and anti-inflammatories into it, to keep it going as long as possible . . . and that's probably why such a high level of carcasses do have detectable levels of diclofenac." Here, poverty emerges explicitly for the first time in this story, but it is a central theme in the unfolding of vulture–human relationships in India today and, as such, a topic to which I will return. In the emergence of vulture-toxic cattle, we encounter the flip side of the proximity and entanglement between people and vultures. While this close association has for a long time been mutually advantageous, it has now become a liability for everyone. Domesticated cattle once provided a great source of carrion for vultures, but this reliance on humans (more accurately, on livestock that they keep) for food may now lead to the vultures' extinction. In other parts of the world where diclofenac and other toxic anti-inflammatories (Cuthbert et al. 2007) are widely used—for example, East Africa—*Gyps* vultures have not been affected by the drug in the same way as their Indian counterparts. This is thought to be the case largely because vultures in East Africa have not struck up a relationship with people similar to that in India, so their diet includes a greater proportion of the carcasses of wild animals (Cunningham, pers. comm.).

Similarly, the entanglement and close proximity of people and vultures in India has become a liability for human communities. In their absence, it has been made all too clear how important a role vultures played through their consumption of the dead, in creating an environment in which the unfolding and flourishing of so much other life could occur. Drawing on Deborah Bird Rose's (2006) work, my position is that in India vultures and others are being drawn into a kind of "double death." For Rose, this concept marks a situation in which life's connectivities are unmade, with disastrous consequences for a whole ecological community. In the story that Rose tells, dingoes are at the heart of this tragedy: having been baited with 1080 poison, they remain toxic even after death so that they poison those who feed on them. Here, death cannot be twisted back into life and instead "starts piling up corpses in the land of the living" (Rose 2006:75). Rose explores the ramifications of this kind of death work among the Aboriginal people of Yarralin (for whom dingo is kin) and their multispecies community.

The use of diclofenac in India is producing a related, but distinct, process of double death. In addition to the creation of an environment in which dead bodies, en masse, fail to nourish but rather poison, the resulting absence of so many vultures has left a vast number of carcasses unscavenged—literally "piling up corpses in the land of the living" (Rose 2006:75). The disappearance of so many members of a species produces what ecologists call a "functional extinction," which may well be followed by an actual extinction in coming years. When vultures no longer inhabit the places and take up the relationships that they once did, the connectivities that make life possible in these places are unmade. As a result, a further "doubling" of death has been set in motion in which all those whose lives and well-being are entangled with vultures are drawn into a process of intensified suffering and death. Here, proximity and connectivity again become a liability, but in a way that highlights some of the inequalities of life, in which poorer nations and, in particular, poorer communities within them are more readily exposed to harm.

As previously noted, in consuming decomposing and sometimes disease-laden bodies, vultures remove sources of potential contamination of soils and waterways while helping to prevent the spread of pathogenic organisms (Houston and Cooper 1975). With a digestive system that routinely processes "foods" that even the most adventurous of others would not touch, vultures are very well positioned to clean up disease threats. For example, this ability may have been connected to the containment of anthrax in India. When an animal dies of anthrax, the spores of the disease often leach into the soil, where they can remain for decades, and can also be spread by wind and in the guts of other animals. In the past, vultures tended to clean off all soft tissue within hours of an animal's death, before the anthrax bacteria have time to form spores and spread (Cunningham, pers. comm.; Houston and Cooper 1975). In the vultures' absence, there are fears that anthrax may become a more significant health problem—especially in the southern states, where the disease remains endemic (Vijaikumar, Devinder, and Karthikeyan 2002). With 70 percent of India's population living in rural communities, and the majority dependent on livestock for their livelihoods, a huge number of people are potentially at risk of infection (Devinder and Karthikeyan 2001).[6]

But anthrax is not the end of this story. In the absence of vultures, it is thought that available cattle carcasses in India may be making room for fast-breeding scavengers like dogs and rats. While there are no accurate figures on street-dog numbers across India, Anil Markandya and his colleagues (2008:198–99) have argued, drawing on Ministry of Agriculture census data, that it seems likely that feral-dog numbers are increasing considerably as a direct result of the decline of vultures. While dogs do consume cattle carcasses, they do not do so with anywhere near the same speed or thoroughness that the vultures once did. As a result, they are not able to provide the same sort of containment of diseases like anthrax, and putrefying carcasses are increasingly left to contaminate waterways and the environment more generally.

In addition, however, large increases in the number of street dogs bring their own problems. According to a study conducted by the Association for the Prevention and Control of Rabies in India (APCRI 2004:44) (and sponsored by the United Nations World Heath Organization), approximately 17 million people are bitten by dogs in India each year; or roughly one person every two seconds. Nationwide, the vast majority of the victims belong to "poor" or "low"-income economic groups (75%), and in rural areas this burden falls even more heavily on these poorer groups (80.3%) (APCRI 2004:25).

While dog attacks are themselves significant, in India dogs are also the primary vector for the transmission of rabies to humans, accounting for approximately 96 percent of all cases (APCRI 2004:44). There are, therefore, fears that the incidence of rabies in India may be beginning to climb as a result of increased dog numbers. It is estimated that 60 percent of the world's rabies deaths already occur in India: approximately 25,000 to 30,000 each year, or one death every thirty minutes (APCRI 2004:44). While vaccines are available for rabies, and they do seem to be reaching many people, the total number of rabies deaths is falling only slightly—perhaps due to a large increase in the number of people being exposed to the disease in recent years (Markandya et al. 2008:199; Menezes 2008:564). In addition to the sheer number of people infected, it should also not be forgotten that contracting the rabies virus leads to a horrific death. According to the British Medical Association's (1995) guide to the disease: "Once clinical symptoms of rabies appear, there is no known cure and the victim is virtually certain to die an agonizing and terrifying death" (13).

Like anthrax and dog attacks, rabies does not have an impact on all social groups evenly. The study conducted by the Association for the Prevention and Control of Rabies in India (APCRI 2004:16) concluded that 87.6 percent of the people killed by rabies in India are from "poor" or "low" socioeconomic groups. In addition, most of these victims were adult men, which the study noted might often place additional economic hardships on families (APCRI 2004:16). It should also be remembered that rabies does not affect only humans, but is often transmitted to a variety of other animals and ultimately leads to painful deaths for many of them, including millions of street dogs.

In addition to threats from environmental contamination and the spread of disease, the mass death of vultures is having economic impacts on some of India's poorest people. These people, often referred to simply as "bone collectors," have made a living gathering the dried bones of cattle and selling them to the fertilizer industry. In the absence of vultures, these bones are now often incompletely scavenged, requiring either extended periods of time before collection or for people to clean the bones themselves (Markandya et al. 2008:195–96).

And so it is clear that while all humans are bound up in ecological relationships inside a multispecies world, we are not all entangled in the same ways. Those who are most directly dependent on vultures will suffer most—in this case, rural and poor communities. But this is by no means an exceptional situation. Drawing on case studies from around the world, the Millennium Ecosystem Assessment (2005:3) highlights the fact that ecosystem degradation and biodiversity loss will often have a disproportionate impact on poor and rural communities (these "two groups" very often being one and the same). These people tend to be directly dependent on their local environments for the provision of "ecosystem services," such as the carcass disposal provided by vultures, or of food and clean drinking water. When these environments are disturbed, these people lack the buffer that markets offer the wealthy, who are often simply able to purchase the good or service from farther afield or buy a substitute. In addition, the Millennium Ecosystem Assessment (2005) notes that "[e]cosystem changes have played a significant role in the emergence or resurgence of several infectious diseases in humans" (27–28). Similarly, when this happens, these communities are likely to lack the basic resources of life, as well as access to medical services, that would treat diseases or prevent them from becoming established in their communities. In this context, vulnerability emerges, at least in part, as a feature of the specific ways in which we are entangled in our multispecies world; some are diffusely entangled, while others are bound up tightly in relationships with *specific* nonhumans and local ecologies (and thus highly exposed to changes or disturbances).[7] In a time of such extreme biodiversity loss and extinctions, these interconnections produce a host of human and nonhuman casualties—lives pulled into amplified patterns of death and suffering.

Paying attention to these changing patterns also helps us to rethink extinction in important ways. In contrast to the simple black-and-white notion of extinction as the singular event of the death of the last individual of a "kind" or a phyletic lineage, Indian vultures remind us that extinction is a far more diffuse and complex phenomenon. Never a sharp, singular event—something that begins, rapidly takes place, and then is over and done with—extinction is by its very nature a slow unraveling of flight ways, of complex ways of life that have been co-produced and delicately interwoven through patterns of sequential and synchronous multispecies relationship (Rose 2012b). The "dull edge of extinction" names this difficult and drawn-out space. As a concept, it is an attempt to articulate a notion of, and focus attention on, extinction as a prolonged and ongoing *process* of change and loss that occurs across multiple registers and in multiple forms both long before and well after this "final" death. To allow the term "extinction" to stand for only the death of the last of a kind is to think within an impoverished notion of "species," a notion that reduces species to specimens, reified representatives of a type in a museum of life, and in so doing ignores the entangled relations that *are* a particular form of life.[8] I am interested in more expansive notions of species and extinction, and so a way of thinking that does not allow a neat separation between "extinction proper" and its various "impacts." This is an approach that refuses this distinction, to focus instead on the undoing of an entangled flight way: a unique, evolved, and evolving way of life whose passing disrupts real lived relations that exceed, cut across, and confuse simplistic categories like the natural and the cultural, the biological and the social, the living and the dead.

Before their numbers began to decline so drastically, vultures were very well adapted to their role as scavengers in Southeast Asia and the Indian subcontinent. As throughout the rest of the world, species of vultures in Asia developed feeding specializations that effectively divided up carcasses, greatly helping to reduce competition when a large number of birds descends on a single carcass. Scientists often classify vulture species into three feeding groups, colorfully referred to as "tearers," "pullers," and "peckers" (Konig 1983). Put simply, "tearer" species have powerful beaks for opening carcasses, while "peckers" have more slender skulls, ideal for the more fiddly work of picking up small particles of food and separating meat from bone. All *Gyps* vultures, the main species being poisoned in India, are "pullers." They are perhaps the most stereotypical vultures, reaching their long, agile, feather-free heads and necks deep into a carcass to pull out the viscera and soft flesh. Because these components of a carcass offer the richest source of food, *Gyps* make up by far the majority of the vulture population in the parts of Africa, Europe, and Asia in which they are found—usually around 90 percent. In India, however, this number has traditionally been closer to 99 percent (Houston 1983:136), perhaps largely because the need for "tearer" species has been undermined by an ongoing entanglement with humans who "open" carcasses when they remove the skin for leather.

The dominance of *Gyps* vultures in India has only added to the tragedy of their extinction. Physiological responses of bird species to anti-inflammatory drugs like diclofenac are generally not well understood. In recent years, however, there has been some suggestion that non-*Gyps* species of vulture and other scavenging birds in India may not be affected by the drug in the same lethal way (Meteyer et al. 2005:714; but see Cuthbert et al. 2007). If this is the case, other species of vulture no longer found in large numbers in India may have survived in the presence of diclofenac. But for the longest time, India has been a land of *Gyps* vultures, knotted together in so many different ways with people inside larger multispecies communities. These processes of knotting, alongside the often painful unknotting that is occurring today, highlight processes of entangled becoming that cut across human/nonhuman and nature/culture binaries to produce a rich multispecies world. In addition, the community I have described is composed of both the living and the dead. The extinction of vultures points to the necessity of a concept and a practice of community that draws in the dead, in which what happens to the dead—how they (and we) are "taken care" of (and by whom) and what contributions they are able to make (and to whom)—is deeply consequential for the health and continuity of others. Dead vultures, dead cattle, dead people, all *matter* in the interactions that produce possibilities of and for life and death.

And so it is not just the living that make, and sometimes unmake, flourishing communities.

But as these relationships have been unraveling in this time of extinctions, those people most closely dependent on vultures have been drawn first into patterns of intensified suffering and death—another form of what Deborah Bird Rose (2006) has called "double death." In this context, it does not seem to be enough to say that "we are all bound up in relationships of dependence in a multispecies world." The brand of holistic ecological philosophy that emphasizes that "everything is connected to everything" will not help us here. Rather, everything is connected to *something*, which is connected to something else (Rose 2008:56). While we may all *ultimately* be connected to one another, the specificity and proximity of connections matter—*who we are bound up with and in what ways.* Life and death happen inside these relationships. And so we have to understand how particular human communities, as well as those of other living beings, are entangled and how these entanglements are implicated in the production of both extinctions and their accompanying patterns of amplified death. This kind of information requires case-specific study and an approach that moves beyond absolute distinctions between humans and nonhumans, the living and the dead.

Equally, it requires an approach to the world in which we "cast our lot for some ways of life [and death] and not others" (Haraway 1997:36). We might argue, for example, that in addition to a doubling of death in some domains, the demise of vultures has actually produced the possibility for new life in other contexts. While vultures may no longer be around to consume carcasses, this tragic situation has made room for the growth of populations of other animals, such as dogs. While many of these dogs live in awful conditions and die painful deaths from rabies and other diseases, their mere presence highlights the fact that food is rarely ever wasted in any absolute sense, a situation that is made clear if we are prepared to shift our focus and acknowledge the broader "company" (Haraway 2003) of species at the table in any, even moderately healthy, ecosystem. Inside rich histories of entangled becoming—without the aid of simplistic ideals like "wilderness," "the natural," or "ecosystemic balance"—it is ultimately impossible to reach simple, black-and-white prescriptions about how ecologies "should be." And so we are required to make a stand for some possible worlds and not others; we are required to begin to take responsibility for the ways in which we help to tie and retie our knotted multispecies worlds (Barad 2007:353–96). How to live well within the always unequal patterns of amplified loss and suffering that are produced here is an issue that can only take on increasing significance as we move ever more deeply into Earth's sixth mass extinction event and a period of growing environmental and climatic change.

Three

URBAN PENGUINS

Stories for Lost Places

> Once penguins have bred at a site it is unusual for them to shift far from there in subsequent years regardless of how much they are disturbed. . . . Little penguins are extremely robust both mentally and physically, and when confronted with human activities, even if adverse, they are unyielding.
>
> CHRIS CHALLIES

There is something remarkable about a shoreline, a place where water meets land and gives rise to a sense of productive confusion between two worlds. For most humans, one of these worlds—the place of earth, of firm land beneath our feet—is home. The other is a place for occasional visits, where we cannot really expect to live our lives, to survive for long periods of time. For penguins, this littoral zone must surely also mark a transition between two worlds, each with its own threats and possibilities. But while penguins are undoubtedly more comfortable, more agile, less vulnerable in the water, they remain utterly tied to the land as well, required to live their lives between these two worlds. Since their distant ancestors abandoned the skies for a life beneath the waves, penguins have maintained a connection to the land, drawn out of the sea each year by their own avian biology and the desire to breed and reproduce.

As biologists Lloyd Davis and Martin Renner (2003) have noted, perhaps if penguins had been marine mammals they would have evolved internal gestation: "the key to the totally aquatic existence of cetaceans and dugongs" (88). But as birds, as egg layers, they need the land to reproduce. Perhaps if they had been marine reptiles like sea turtles, they

A Little Penguin coming ashore at night. (Richard Fisher; CC BY 2.0)

would have been able to come up onto the beach in a single night, deposit their eggs, and then disappear back into the ocean. But as birds, penguins are homotherms, and so "their eggs must be kept warm for development to take place" (Davis and Renner 2003:88). Finally, unlike many other seabirds—for example, the albatrosses discussed in chapter 1—penguins cannot fly great distances between their breeding and feeding places. Thus in addition to all the other requirements of a breeding site, penguins need a place where there is close access to a steady supply of food (Davis and Renner 2003:88). And so it is as a result of the unique biology of this bird that lives much of its life under the water, that penguins have developed their own particular relationships with the land—or, more accurately, with the very few specific places that they come ashore each year to reproduce. For, as we will see, these are *specific* places, not at all interchangeable, but deeply storied, carrying the past experiences of individuals and the generations before them.

The subject of this chapter is one such penguin place: a thin band of rocky foreshore at Manly, just inside the mouth of Sydney Harbour, Australia. Here, among the sandstone crevices, under waterfront homes and boats, and in a range of other strange locations nests a tiny colony of Little Penguins (*Eudyptula minor*). Each year, these penguins return to this place to breed and molt. Members of the world's smallest penguin species, they stand roughly 1 foot (30 cm) tall and weigh around 2 pounds (1 kg). While others of their kind continue to form large colonies on islands off the coast of Australia, the same is not true for the mainland. Only fifty years ago, Little Penguins could be found nesting at several sites in the harbor, as well as at various locations farther south along the east coast of Australia. In recent years, however, all these colonies have been lost. The tiny Manly colony, composed of around sixty breeding pairs, is now thought to be one of the last three on the Australian mainland and the last in the state of New South Wales.

As a result, in 1997 the colony was listed as an "endangered population" under the state's Threatened Species Conservation Act (1995). Despite its protected status, however, this colony continues to suffer from a range of threats. Perhaps most significantly, the shoreline that they rely on for nesting is being transformed and lost to them. While some of the penguins nest on land that is part of Sydney Harbour National Park (North Head), many of them are to be found on a thin strip of the shore lined by waterfront homes that often have made penguins' presence either impossible or much more difficult (for example, through increased noise, light, and disturbance, including predation by pets).

This loss of nesting places was made all too apparent toward the end of the 1980s when the owner of a little house on Manly Point built a seawall along the edge of his property. He claimed that despite his efforts to encourage them, penguins had not been present on his land for several years. Almost immediately, however, the wall attracted the attention of local activists who claimed that penguins had been deliberately discouraged from the site: the wall had blocked their access to their nesting places, and after the persistent little birds had found a way through the wall—by way of a drainage pipe—even this path had been blocked. But penguins do not give up easily when obstacles are placed between them and their burrows. Even if the place has not been used in any one year, Little Penguins know where they bred last, and in most cases they will attempt to return: in the language of biology, Little Penguins possess a strong "site fidelity." After this particular seawall was built at Manly Point, locals reported penguins coming ashore farther along the coast, making their way across a small beach, up a flight of stairs—which is no small feat for a Little Penguin—along the street, down another flight of stairs, and under the house that they had been so unceremoniously evicted from (NPWS 2000:12). But the land is not a safe place for penguins, and throughout the breeding season several are thought to have been hit by cars or taken by dogs (NPWS 2000:24).

While this seawall attracted a great deal of opposition, it was not the first such structure in the area, and it will not be the last. As I stood on the foreshore of Manly Point in early 2010, this little seawall seemed to me to be remarkably unremarkable. Far from being one of a kind, the wall actually filled what would have been a tiny gap in an uninterrupted patchwork of seawalls that continued on steadily in both directions.[1] Almost all the houses along this stretch of the harbor have similar walls, separating private land from the water in a way that undoes any of the ambiguity of—and, with it, penguins' easy access to—the shoreline. But seawalls have been being built here for almost as long as houses have.[2] Since at least the early twentieth century, those homeowners who could afford to have built seawalls: to claim land from the harbor for a backyard or, more commonly, to moor a boat or build a swimming pool. While sandy spots have been prized as beaches, the far more common rocky shoreline—once used for penguin breeding—has very frequently been covered over by a wall. Neither is Manly exceptional in this regard; an estimated 50 percent of the shoreline of Sydney Harbour, hundreds of miles in length, is now composed of seawalls or other built structures (Chapman and Bulleri 2003). But these seawalls are just one small part of the problem for penguins in this urban environment that each year becomes a little more built up, a little more noisy, a little more dangerous.

And yet, year after year, the penguins keep returning.

I am captivated and unsettled by a singular image, and it is to this image that this chapter responds. It is the image of a penguin returning to a burrow, to a breeding place, that is no longer there or has been transformed so dramatically that it is no longer habitable. Of houses and swimming pools lining the shore of a harbor: people pulled to be closer to "nature," but in a way that makes life for penguins and other harbor dwellers impossible. But this story of returning to lost places is not solely about penguins or even coastlines.[3] The returning penguin could just as easily be any number of other migratory or nomadic birds, a sea turtle, or even a seal looking for a haul-out spot on the shore. The beach could be a mangrove, a wetland, a tidal flat, or a range of other settings. All over the world, animals are drawn to return faithfully to places that no longer exist. Often they return to these places to breed, and often this is when they are at their most vulnerable—as with penguins returning to dry land. All these animals are in the background of this multispecies story about penguins and people along an urban shoreline in Australia. While each of them has their own unique story, my hope is that this singular account may open up some general lines of inquiry about those many others that are fatally tied to disappearing or lost places.

STORIED-PLACES IN ANIMAL WORLDS

Places are not at all abstract or interchangeable; rather, they are nested and interwoven with layers of attention and meaning.[4] In this context, a place is more than the "raw" biophysical landscape: in Edward Casey's

(1996) terms, "a place is not a mere patch of ground, a bare stretch of earth, a sedentary set of stones" (26). Rather, "place" must be understood as a material-discursive phenomenon. Denis Byrne, Heather Goodall, and Allison Cadzow (2013) make this point succinctly when they note that "humans make places out of spaces not just by physically altering them but also via the social and mental process of making them meaningful" (26).[5] This understanding of place highlights its "storied" nature: the way in which places are interwoven with and embedded in broader histories and systems of meaning through ongoing, embodied, and inter-subjective practices of "place-making" (Byrne, Goodall, and Cadzow 2013; Casey 1996:2001; Malpas 2001).

But it is not only humans who have the capacity to "make" places in this way. Many nonhuman animals are also "generators of meaning" (Lestel and Rugemer 2008:9). As Eduardo Kohn (2007) notes, "[T]he biological world is constituted by the ways in which myriad beings—human and nonhuman—perceive and represent their surroundings" (5). In a host of different ways—differences often covered over by that abstraction called "the animal" (Derrida 2008; Peacock 2009)—these other living beings constitute their worlds as places richly meaningful, historical, and storied. As Barbara Noske (1989) puts it: "[I]t is not just human subjects who socially and collectively construct their world but . . . animal subjects may do so too. The animal constructs are likely to be markedly different from ours but may be no less real" (157–58).

The work of the early-twentieth-century Estonian biologist Jakob von Uexküll ([1934, 1940] 2010) is central in the history of scientific attempts to take seriously the meaning-making abilities of various nonhumans. In this context, von Uexküll described an organism's *Umwelt*: the experiential world that a being inhabits, its perceptual surrounds. Living beings perceive their surrounds differently as a result of their particular embodiment—sight, smell, and other senses—as well as through their specific needs, desires, and own life histories. As a result, they occupy different *Umwelten*. In taking up this approach, von Uexküll rejected the notion of animals as "physico-chemical machines" whose movement through the world can be explained by simple "instinctual, mechanical responses" (Buchanan 2008:7, 31). Instead, he described a "biology of subjects" in which, in Brett Buchanan's (2008) words, organisms are understood to "actively interpret their surroundings as replete with meaningful signs. They are not merely passive instruments or message bearers, but actively engaged in the creation of a significant environment" (28, 31–32).

The experiential worlds of these animal subjects are invariably difficult, and to some extent impossible, for us to grasp, in part at least because our access must occur through our specifically hominid embodiment. For example, the acoustically mapped worlds of dolphins using echolocation are probably impossible for humans to comprehend on a range of levels (Noske 1989:159). Similarly, as Dorion Sagan (2010) notes when discussing the capacity of blue whales to communicate by song over vast distances: "They may, in their giant *Umwelten,* have fabulous multisensory pictures of major portions of the ocean, images that, even if we had direct access to them, we couldn't process, because our brains are too small" (23). But this lack of access on our part should not stop us from attempting to describe the kinds of worlds that these beings inhabit—however imperfectly—and it certainly should not lead us to deny that they inhabit meaningful worlds at all.

And so my focus in this chapter is the experiential world of Manly's Little Penguins. In particular, I am interested in their relationships with the shoreline that provides their nesting places. As is perhaps clear, I am interested in more than the ecological requirements of appropriate nesting "habitat." Specifically, I am interested in how these penguins "story" these places, how breeding places are rendered historical and meaningful to penguins. In using the term "story," my claim is not that penguins know and "do" places in the same way that people do, but simply that their relationships with place might also be productively understood through this conceptual lens. While a great deal has been written on the subject of human "place-making" in recent decades, less attention has been paid to the complex and diverse ways in which nonhumans might also story their places.

The notion of "story" at work here is a basic but foundational one: the ability to engage with happenings in the world as sequential and meaningful events. The environmental historian William Cronon (1992) has drawn an instructive distinction between "narrative" and "chronology" (1351). He notes that a chronology is a simple listing of events in their order of occurrence. In contrast, a story, or narrative, weaves those events together in a way that generates context and meaning. Connection and relationship are central to narrative. Events do not just happen one after the other in a random sequence; rather, they are connected to one another, and they affect or cause one another in a range of ways. Story is about the weaving of those connections, either in the recounting of events (storytelling) or simply in one's own "storied experience" of the world.

Although cast in a different language, it is precisely this distinction between chronology and narrative that is at issue in many discussions of "animal mind." In her account of the history and philosophy of ethology, Eileen Crist (1999) distinguishes between an experience of events in the world as "sequentially connected" or merely "serially placed" (170). She notes that many of the more reductive and impoverished approaches to behavior and cognition have tended to present animals as though they have no *cohesive* experience of the world. Rather, their actions are understood as things that happen *to* them—driven largely by "instinct" or "stimulus–response mechanisms"—not the result of any overarching reasoning or understanding on their part of the way in which these actions or broader events in the world connect to causes and outcomes. The result is an image of animal life as a fractured and disjointed set of "serially placed" experiences, occurring one after the other, but lacking any meaningful organization for the animal itself (a "chronology," not a "narrative," in Cronon's [1992:1351] terms).

But as work in ethology, cognitive science, and related fields has made increasingly clear over the past several decades, this is not the way that many animals (in addition to human animals) experience their worlds (Allen and Bekoff 1999; Goodenough, McGuire, and Jakob 2010; Wynne 2002). As Darwin (1871) noted far too long ago for it to remain a surprise, evolution demands an understanding of humans' mental and emotional capacities as continuous with those of the other members of the animal kingdom: "[T]he difference in mind between man and the higher animals, great as it is, certainly is one of degree and not of kind" (101).[6] It was with this understanding in mind that Darwin, and many others since him, presented an image of animal experiences as "sequentially connected": a world in which behaviors and events are placed in a meaningful context by virtue of their relationships with one another. As Crist (1999) notes, it is precisely because of their dwelling within a "cohesive" and "continuous" time—one that enables the experience of events and actions as significantly connected to one another—that animals inhabit "a meaningful world" (170), what I am here terming a "storied experience."

And so this chapter is an attempt to draw recent work on the biology, ecology, and ethology of Little Penguins into conversation with previously human-centered work in the humanities and social sciences on story and place. In so doing, we are able to explore what it might mean to take seriously the very different ways in which various nonhumans understand their places and render them meaningful. What kinds of competing claims to places might emerge from an account of storied-places as genuinely and pervasively *multispecies* achievements? More concretely, what would it mean to take seriously the way in which the penguins of Manly story their specific breeding places?

PENGUINS-STORIES-PLACES

Little penguins are philopatric, a term that literally means "love of one's home" and in biology describes a process in which an animal returns to its place of birth or hatching to reproduce. It is not clear how, or precisely when, this attachment to a natal place develops. For roughly the past half century, curious biologists have moved seabird hatchlings of different ages between colonies to see which site they would return to. What seems to have emerged out of all this geographical confusion is that philopatric attachment develops at some point between hatching and fledging. If chicks are moved before this magical time, they can be expected to return to the new site to breed in roughly the same numbers as birds that actually hatched there (Serventy et al. 1989). Chris Challies's (pers. comm.) detailed research with Little Penguins in New Zealand over the past thirty years indicates that translocation has to occur before a chick goes to sea for the first time—which happens at around fifty-five days of life.[7]

However it develops, this strong philopatry means that when it comes time to breed, most Little Penguins find their way back to their natal places, often traveling long distances to do so. In fact, even before they are sexually mature, many penguins return to these sites and come ashore at night during the breeding season (Challies, pers. comm.). While a few individuals do opt to visit and eventually breed in a colony other than the one in which they hatched, once they have bred in a place for the first time—irrespective of whether they hatched there—penguins have a very high degree of fidelity to that place (site fidelity). So strong is this connection that it has on occasion been taken advantage of by biologists trying to remove Little Penguins from an unsafe area. For example, in 1995 penguins were translocated hundreds of miles from their breeding area during an oil spill off the coast of Tasmania (after the cargo ship *Iron Baron* ran aground). The long period required for the return swim provided time to begin the cleanup of the area, while minimizing the risks of disease and stress to the penguins that might have resulted from their being held in captivity (Hull et al. 1998).

Interestingly, this fidelity to a breeding site is often very spatially specific, with Little Penguins returning not only to the same general area, but usually to the same burrow or nest each year (Rogers and Knight 2006). But this fidelity is also not absolute. Several of the detailed studies on Little Penguin colonies in Australia and New Zealand have found that birds are significantly more likely to change nests if they were unsuccessful in their previous breeding attempt (Bull 2000; Johannesen, Perriman, and Steen 2002:245; Reilly and Cullen 1981:81). In addition, a study by Edda Johannesen and colleagues (2002:245) suggests that this willingness to change nests may be, to some extent, dependent on the availability of a superior nesting site nearby.[8]

Various explanations have been offered for this site fidelity, including the suggestion that it enables birds to retain high-quality nests and ones with which they are familiar. It may also minimize the time required to prepare a nest or burrow and increase a penguin's chances of reunion with a past mate (in addition to site fidelity, Little Penguins display fidelity to their breeding partners, perhaps especially partners with whom they have bred successfully in the past [Rogers and Knight 2006]). It is also possible that the other penguins that live in a colony may be of significance to any individual, with some studies showing that Little Penguins may preferentially associate with specific others at sea—a behavior that, among other things, may increase their chance of success in the all-important search for food (Daniel et al. 2007).[9]

Wherever they go to breed, however, for Little Penguins the presence of the colony is all important. As with many other seabirds, is seems that Little Penguins will not nest in a place where other birds of the same species (conspecifics) are not present. In this context, the sight and sound of other birds seems to play an important role in penguins coming ashore—indeed, even after establishing nests in an area, penguins usually gather in a group out at sea (called a "raft") and come in to the beach as a group. If juvenile birds return to their natal place and find it abandoned, it is unlikely that they will attempt to breed there. The importance of the presence of conspecifics has been reaffirmed in experiments in which recorded bird sounds and bird models have been used, with some success, to lure juvenile penguins and other seabirds to potential new breeding sites (Gummer 2003; Podolsky 1990).

These comments point to a general pattern of terrestrial behavior for Little Penguins. While there is likely a great deal of variability among individual penguins in breeding and site fidelity, in general terms it seems clear that a range of complex factors are at work in the formation of any penguin's relationship with its nesting place. Places emerge here as storied landscapes: remembered, reinterpreted, and imbued with a changing value and significance through the course of a penguin's life. Initially, there is a pull to return to a natal site; that pull is then influenced by some specific changes in the site, especially with reference to other penguins' presence, as well as the individual's own experiences (perhaps, in particular, past breeding success or failure).

But these stories are not just layered over an existing world. Rather, they emerge from and influence the way in which places and those that dwell in them come to be. A storied shoreline and a colony of penguins emerge here in a process of entangled becoming, where none of the relata (at least in their current forms) preexist their relationships (Barad 2007). While only some shorelines offer an appropriate environment for nesting penguins, once these birds settle in, they physically alter the terrain in a range of ways, especially through ongoing burrowing, breeding, hunting, excreting, and more. In fact, guano from penguins and other seabirds is often a particularly important component of the nutrient cycles of coastal regions and small islands, depositing much-needed nitrogen and other nutrients (Gill 2012; Muller-Schwarze 1984:26; Stearns and Stearns 1999:9–10).

But at the same time that penguins change the shoreline, they themselves are remade through this relationship. Over the years, perhaps centuries or longer, the penguins in the Manly colony have adapted their breed-

A Little Penguin. (Kenneth Fairfax; CC BY 2.0)

ing behavior to the unique local environment: in the absence of tussock grass and sandy soils, into which in other places penguins dig burrows, members of this colony, located in sandstone country, have had to utilize primarily rock crevices for their burrows (Bourne and Klomp 2004:131). In recent years, they have been required to again adapt their breeding behavior, this time to make use of, as well as gain protection from, a changing urban environment. Local penguins are sometimes to be found nesting in the dark and dry places underneath houses, sheds, boats, and other structures and vehicles. As Julie Bourne and Nicholas Klomp (2004) note: "These modifications to their nesting behavior have enabled Little Penguins to persist in the densely urbanized environment of Sydney Harbour" (131).

But, as we have seen, this is the only spot in a harbor once rich with penguin life that survival has been possible at all: a single, tiny colony—likely greatly reduced in numbers—and even now only just hanging on. Thinking through the lens of storied-places enables us to appreciate some of the ways in which penguins, places, and the stories that connect them are all at stake in one another, all reshaped through ongoing patterns of attachment and relationship.

LOVED AND LOST PLACES

We do not really know how long the penguins have been returning to Manly, although they are widely thought to have been there since well before British settlement of the colony of New South Wales. By the mid-nineteenth century, not too long after this settlement, Manly's beaches had already become an important recreational site for Sydney's residents. Initially accessed by ferry, the seaside destination boasted that it was "seven miles from Sydney, and a thousand miles from care" (Curby 2001). Over the intervening decades as Sydney expanded, Manly was slowly subsumed within the greater city limits, giving rise to a variety of new problems for penguins. The earliest documentary evidence of penguins' presence in the area is from 1912, and it mentions only a large group of penguins arriving in Sydney Harbour near Manly, not any breeding activity (*Evening News* 1912). Occasional newspaper articles from the 1930s and 1940s refer to penguins in the harbor (*Sydney Morning Herald* 1936, 1948), and one notes that they were breeding at Quarantine Point at this time (*Sydney Morning Herald* 1931), where a small number of them can still be found.

An article from the mid-1950s again mentions breeding penguins around Manly (*Sunday Telegraph* 1954). From this piece, it is clear that the colony was once considerably larger than it is today. Unfortunately, this fact is communicated to us through a report that more than 300 penguins were shot dead in one night on a single beach in an act of "vandalism." While the newspaper reporter deplores this act of violence, it is clear that many other Sydney-siders in the 1950s did not enjoy sharing their space with Little Penguins (in a similar act of violence two months later, thirty penguins were killed by "hoodlums" in Terrigal Beach, north of Sydney [*Sun-Herald* 1954]). An article in the *Australian Women's Weekly* in December 1956 highlights this situation clearly in a short photographic spread: images from Narrabeen, about 6 miles (10 km) up the coast from Manly, show residents boarding up the gaps underneath their houses to prevent penguins from nesting in these spaces. It seems to have been primarily their noise—described as a "nightly 3 AM party on the beach"—that caused the most difficulty for the locals. The article notes that "as daytime guests they're welcome, but at nightfall they head down to the sea for food—making noises that keep everyone else awake, too." It ends by conceding that, as the penguins are a protected species, the residents can work only to deter their presence and must "resign themselves to a trying time while the penguins . . . are in charge" (*Australian Women's Weekly* 1956:22–23).[10]

A lot has changed in the intervening years. Like almost all the other penguin colonies on the mainland of New South Wales, the penguin colony of Narrabeen is no more. As with the builder of the seawall mentioned earlier, some residents of Manly continue the long tradition of discouraging protected penguins from taking up residence on "their" property. On the whole, however, these *deliberate* efforts to discourage the penguins' presence are probably few in number and are far less significant than the widespread loss of breeding sites that has resulted from the relentless densification of the area. The number of human visitors to and residents of Manly, one of Australia's most iconic beachside suburbs, has steadily increased over the past several decades and can only be expected to continue to do so. Steadily rising land prices have ensured that blocks of land have been subdivided, and the human footprint has crept closer to the water's edge. Even the increasingly popular harbor-side swimming pools in this area have often been fatal for Little Penguins, with several being trapped and drowned in them each year.

It is into this density of human habitation that the Little Penguins must return each year. Sometime around July, they enter the mouth of the harbor. Looping around the bottom of North Head, they make their way to the shoreline and their burrows beyond. For roughly the next eight months, they move back and forth between burrow and water, usually under cover of darkness. For the first few months, they are occupied with breeding: nest preparation, copulation, incubation, and then chick protection and feeding. When the chicks are finally ready to fledge, the adults return to the sea for a few weeks of fattening up, before coming back ashore for their annual molt. During this molting period of roughly two weeks, they are on a starvation diet, unable to go to sea to feed without the protective warmth of their feathers.

As any, even casual, observer of penguins knows, they are not ideally suited to this terrestrial portion of their lives. While the water is certainly not free from danger for penguins—in the form of marine predators, as well as boats, fishing lines, and other forms of pollution—on land their slow and awkward waddle makes them easy targets for predators, including, in Australia, dogs, foxes, cats, birds of prey, and occasionally people. This situation is only made worse by the long periods of time out of the water required for breeding and molting. In both cases, a dry and secure burrow is a necessity. It is, therefore, at their most vulnerable times that penguins make their way onto the shore—returning to the places where they were hatched, the places where they have perhaps hatched their own young, the places that should be safe at this most precarious of times.

The ongoing urban development in the area impinges on the colony in a range of ways. The increased number of people living near or along the shore means more human presence at the water's edge, more dogs moving around to disturb and attack penguins, as well as increased light and noise along the water. In addition to direct penguin mortality, all this activity functions as a strong deterrent for penguin parents to return to their burrows to feed their chicks—likely resulting in reduced chick health and survival (NPWS 2000:24). Finally, urban densification, of course, also leads to the *direct* loss of nesting sites in the area, as suitable sites are transformed into swimming pools or houses, or as penguins' access to them is simply blocked by a wall or another structure. All these pressures have worked together to significantly reduce the availability of breeding sites. According to the National Parks and Wildlife Service, the organization tasked with conserving this endangered population of penguins, the loss of suitable breeding habitat is a "major threat" to this colony and now also "seems to be the main factor limiting the[ir] distribution" (NPWS 2002a:13).

And so it is here, in this narrow, and highly valued, littoral zone—a rocky foreshore squeezed between the water and a growing number of buildings—that penguins and people have been thrust together. This is a space that is prized by its human inhabitants primarily for its harbor views and water access. But the unwillingness or inability of the local community to genuinely hold open space for penguins highlights a disturbing and all-too-frequent dimension of life along the coast, in Australia and elsewhere. While Sydney Harbour is highly valued by people, it is imagined first and foremost as a site of human amenity—calm waters lapping against a bank (or, more commonly, a seawall)—not as a place that is vitally important to the lives of a variety of nonhuman others that make their homes on the watery edge of the city (NPWS 2002b).

Underlying this tangible loss of penguin nesting sites is a discursive framing that effaces the penguins' presence and refuses to recognize any significant claim by them on this place. This situation is readily apparent in the language of "unwanted guests" and "reclaimed" shoreline. The "re-" in "reclaiming," of course, implies a prior ownership or entitlement to something. In this context, the shoreline is not appropriated, or taken, so much as it is "returned" to its rightful owner. But there is another important dimension to reclamation, that of "improvement." In its application to land and resources, the term "reclaiming" is usually used to indicate a redirection or transformation of something that is otherwise wasted. The *Oxford English Dictionary* makes this point succinctly, defining "reclamation" as "[t]he conversion of wasteland, esp. land previously under water, into land fit for cultivation or construction." In the context of Sydney Harbour, the prior configuration of this place—its rocky shoreline providing safe burrows for penguins and access to the land beyond—is cast as wasteful and irrelevant. The only meaningful use is human use: an extended backyard or an "infinity pool" hanging out over the harbor.

There is a similar dynamic at work in the positioning of penguins as "guests," especially of the unwanted variety, as in Narrabeen. Here, humans are cast as the rightful inhabitants of this place. The shoreline is "our land," and we may or may not extend hospitality to others. This framing of claims to the shoreline is "infected with a selective forgetting" that renders invisible a prior "taking" (Diprose 2002): our various arrivals and radical transformations of a shoreline (and, indeed, a country) in a way that covers over and undermines the ongoing presence of others and their claims to these places. As Rosalyn Diprose (2002) notes, acts of giving are often, perhaps always, premised on prior takings and enclosures, many of which are unacknowledged or deliberately rendered invisible. There are obvious connections here to the treatment of indigenous people and more recent immigrants and refugees, who are often similarly positioned as lacking any legitimate claim to places in Australia.

My primary interest, however, is in the effacement of the presence of penguins and a range of other species that once lived in and along the harbor. Excluded in the past—through either neglect or deliberate, and often violent, action—some of these animals are now making their way back or having their continued presence more actively acknowledged and supported by some people. But on whose terms is this happening; who is required to make room for whom? Equally important, how is the language of "the guest," which is often used in reference to these animals, complicit in erasing a past displacement and claim while also creating an unstable future, where residence can only ever be temporary and on "our" terms (Thomson 2007; van Dooren and Rose 2012)? In short, the questions is: Who are we to welcome penguins to this shoreline as guests? As Jacques Derrida (1999) notes: "To dare to say welcome is perhaps to insinuate that one is at home here, that one knows what it means to be at home, and that at home one receives, invites, or offers hospitality, thus appropriating for oneself a place to *welcome* [*accueillir*] the other, or, worse, *welcoming* the other in order to appropriate for oneself a place" (15). In the case of Little Penguins and many other nonhuman inhabitants of urban places, however, the "guests" are not even welcome—and so their presence and their claim are doubly effaced.

SHELTERING GENERATIONS

In this context, paying attention to penguins is about honing our skills at listening for alternative and often "unspoken" stories; it is about learning an appreciation for more-than-human practices of meaning and place-making in a disappearing world. I agree with William Cronon (1992) that "narratives remain our chief moral compass in the world. Because we use them to motivate and explain our actions, the stories we tell change the way we act in the world" (1375). But living well with others can never be about just learning to *tell* new stories; it must also involve learning new kinds of attentiveness to the stories of others—even if they are unspoken or are told in other-than-human languages. In taking up this approach, I am explicitly rejecting the common notion that narrative is an essentially, and perhaps constitutively, human capacity (Kearney 2002:3). Cronon (1992) seems to hold this view when he asserts that narrative is "a peculiarly human way of organizing reality" (1367). But experiencing beings like penguins "represent" the world to themselves, too (Kohn 2007:5); they do not just take in sensory data as unfiltered and meaningless phenomena, but weave meaning out of experiences (van Dooren and Rose 2012), so that they, like humans, "inhabit an endlessly storied world"

(Cronon 1992:1368). These diverse multispecies perspectives play havoc with the simple notion that "nature is silent," an un-storied landscape awaiting the human inscription of meaning.[11]

In being attentive to the stories of penguins and others, we help to challenge the closure of human-centric narratives, narratives that along our coasts all too often cover over nonhuman needs and voices. In so doing, we also begin to undermine the obviousness of human understandings and meanings in these shared places, a project that is an essential first step toward ethical relationships. As Val Plumwood (2002) succinctly put it: "Recognising earth others as fellow agents and narrative subjects is crucial for all ethical, collaborative, communicative and mutualistic projects" (175).

While human understandings of the area of Manly tend to focus on the larger city of Sydney, which spreads out from the coastline, penguins surely inhabit an entirely different geography and possess an entirely different sense of what this place means and of the way in which it fits into and relates to the places around it. They likely have little sense of the city that lies beyond the foreshore, but instead know it as a thin strip of land connected to an ocean and a harbor that is rich in the fish and squid so necessary for successful breeding. It is a rocky place, one that provides unconventional but solid burrows for protection from predators. But, perhaps more than any of the current advantages or disadvantages that it offers, Manly is a place intimately known, used for generations, and as little as we understand about the impulses or mechanics of avian fidelity, it is clear that it is a place that calls out in some way to be returned to.

As previously noted, this is a relationship that goes well beyond what we ordinarily mean by "habitat," a concept that usually refers to a purely physical set of features and relationships. In this context, habitat emerges as a largely interchangeable place, as is clear in the *Oxford English Dictionary* definition of the term, which notes that it is "chiefly used to indicate the *kind* of locality, as the sea-shore, rocky cliffs, chalk hills, or the like" (emphasis added). As long as a locality possesses the requisite ecological and biological characteristics, it will be "suitable habitat" for a particular species. For example, in the case of Little Penguins, breeding habitat cannot be too warm (because they overheat easily on land); it must be close to a suitable food supply (because they cannot swim too great a distance while incubating eggs and guarding chicks); it must provide dry and secure burrows within easy reach of the water; and it must be home to a significant number of other Little Penguins.

While all these characteristics are important, as we have seen they are far from being all that there is to the ways in which Little Penguins know and value their breeding places. Any piece of land that meets these requirements is not just as good as any other. Only one colony is "home" and, within it, likely only one burrow. More than the sum of their ecological parts, these places carry penguin histories and stories. In focusing exclusively on "habitat" in accounts of penguin breeding places, we provide a framework of thought in which it is far easier to deny, or conveniently forget, both the real significance of penguins' relationships with these *particular* places and the fact that penguins inhabit their own richly meaningful and storied worlds. It is precisely this inability or unwillingness to recognize penguins' relationships with local places as significant—as meaningful and vital—that enables us to so blithely evict them from a shoreline. In this context, what has been usurped is not a home, not a meaningful and important place, but a piece of interchangeable "habitat." And so the inability or refusal to recognize how penguins relate to *particular* places undermines the significance of their relationships to these places and, in so doing undermines the importance of the claim that they make on them. But penguins do not occupy "habitats." Rather, they inhabit experiential worlds in which a burrow might meaningfully be understood as a "home."

Through this unique relationship, these particular places carry and shelter the possibility of the continuity of penguin generations. In this context, the loss of these places will have a profound impact on the possibilities of future generations. As a breeding area becomes unsafe or disappears, penguins are unlikely to evaluate the site as no longer appropriate and simply move on. As noted by Chris Challies (pers. comm.) in the epigraph to this chapter, Little Penguins tend to return to and remain at their general nesting site after they have bred there for the first time: "Little penguins are extremely robust both mentally and physically, and when confronted with human activities, even if adverse, they are unyielding."

While there have been a very few examples of other penguin species that may have moved their breeding sites when threats became too great, this is at best a very infrequent behavior (Gummer 2003:17). Instead, they stick it out, perhaps changing burrows if a mate is killed or a site becomes too full, but otherwise being largely unresponsive to serious changes. And so as places become increasingly degraded, most penguins will return and be killed. Or, at best, be unable to successfully reproduce at the levels that are necessary to ensure the continuity of the population.

The extreme form of attachment that penguins exhibit may lead some people to regard them as dim or even unthinking. But the intelligence of penguins, like that of all animals (including ourselves), is a product of a long evolutionary history that has determined the kinds of phenomena and environmental change that a being has to be sensitive to. While, as we have seen, penguins are responsive to a variety of factors in their evaluation of a breeding site, the kinds of large-scale change now commonly brought about by people have often not registered for them—just as they have not for albatrosses and many other colonial seabird species (chap. 1). And so we need to develop ways of thinking about animals and their experiential worlds that are respectful of these diverse sensitivities, without always regarding seemingly illogical behavior (from a human perspective) as a sign of stupidity or, worse, the complete absence of mind and any meaningful relationship with the world. This is not a question of more or less intelligence, but of a "diversity of sensitivities," each appropriate to the life way of a given species. Along similar lines, Roberto Marchesini has suggested that we might understand these as "multiple intelligences" (Bussolini 2013). In the case of Little Penguins, the important point is that they simply are not sensitive to many of the perilous changes now occurring in the places where they nest.

In an important sense, we might understand these storied-places as intergenerational gifts. While they take form through the lives and experiences of individual birds, they are not the product of any single penguin. Rather, the cumulative experiences of a penguin's forebears are passed between generations when a hatchling inherits its nesting site from its parents. Here, a vital connection with a specific place that has been found to be productive and safe is maintained and passed down. If all goes well, it will ultimately be passed down again, perhaps in a slightly different form. In this context, these storied-places are themselves deeply entangled in the evolution of the species and the history of specific colonies: both the ability to understand and relate to places in this way (through the evolved capacity of philopatry) and the specific places identified to new generations through their hatching form an important inheritance.

This intergenerational gifting highlights another important aspect of the way in which penguins and their nesting places are enfolded into each other. In both their individual lives and the life of the colony and species—stretched across evolutionary time frames—penguins and places are at stake in each other, unable to be neatly teased apart. Penguin reproduction, like that of all living things, is never simply about the transmission of genes between generations. Genes sit within cells; through ontogenesis, cells slowly become bodies. But in avian worlds, this happens only if those bodies are cared for: if eggs are laid, incubated, and hatched and if chicks are fed, reared, and fledged (chap. 1). In addition, Manly's penguins remind us that this work does not happen in a vacuum, nor does it happen in any old environment. Alongside inheriting genes, organisms also inherit environments (at multiple scales and with a range of significances). In this context, the shoreline is part of what Meredith West and Andrew King (1987) have called an "ontogenetic niche": the broader biophysical environment of cells, bodies, eggshells, and external environments like shorelines that make reproduction possible at all. As Susan Oyama (2000) notes, with particular relevance for those who tend toward genetically reductive accounts of reproduction: "[T]he niche that the genes 'are inside of' is an indispensable bridge between generations" (62).[12]

Storied nesting places are at the heart of the Little Penguin's ontogenetic niche. They are a key part of the inheritance that makes the ongoing life of the colony and the species possible at all. Again, penguins and their places cannot easily be teased apart. The nature of their entanglement means that destroying these places and excluding penguins from them leads inexorably not just to the loss of a few individuals, or a single generation, but to the loss of the possibility of the continuity of generations as such. Whole family lines will be ended here. This is the work that fuels species extinctions, that undermines the ability of populations to sustain themselves into the future, to gift both life and a successful way of life to the next generation.

But, as we have seen, Manly's penguins continue to be threatened in precisely this way. Rich practices of more-than-human gifting are overridden or ignored by a "taking of places" that fails to recognize the vitality and significance of other species' practices of place-making—of individual, collective, and intergenerational attachment and inheritance. While it might be hard to fully comprehend their significance, it is ultimately in little unassuming places like this short stretch of urban coastline that the ongoing possibility of generations is sheltered.

Taking penguin stories seriously opens our world into an attentiveness to this very particular and consequential relationship between a bird and its place. It opens up new possibilities for understanding the loss of coastal places (and many others as well). This is an approach that takes seriously Donna Haraway's (2008) injunction to genuinely get to know the organisms that we philosophize about: "Caring means becoming subject to the unsettling obligation of curiosity, which requires knowing more at the end of the day than at the beginning" (36). Knowing more matters, not least because it can and does enable us to see differently, and so to be drawn into new kinds of relationships, new ethical obligations. In this context, getting to know penguins and their philopatric ways must give rise to an appreciation of the actual ethical weight of our destructive actions in littoral places. An appreciation of the entangled intergenerational fates of penguins and their storied-places makes clear that destroying and usurping these places is very definitely "extinction work"—perhaps not today or tomorrow, but certainly in the all-too-immediate future. In this context, it becomes hard to overstate the significance of these places and the wantonness and severity of the act of quietly destroying them as though there were plenty more available coastline elsewhere.

And yet all over the world, other birds and animals are also returning to these kinds of lost places, to inter-generationally gifted places that are changed or that no longer exist at all—to tidal flats where land has been reclaimed and buildings now spill out into the water, to beaches now covered by people or bathed in city lights (Oldland et al. 2009). In other places, it is "coastal armoring"—seawalls, revetments, and other mechanisms constructed to protect houses built right along the water—that is causing problems, as with nesting sea turtles on Florida's beaches (a state that hosts 95% of all sea-turtle nesting in the continental United States [Mosier and Witherington 2001]). In yet other places, demands for land for agriculture and industry are driving this loss. The tidal flats of the Yellow Sea offer a tragic example of this situation (MacKinnon, Verkuil, and

Murray 2012). These areas are a vital staging ground for at least 2 million birds on their annual migrations—providing food and rest halfway through a mammoth migration of sometimes well over 6,000 miles (10,000 km) each way. And yet when these birds arrive, these areas are now often gone: almost 50 percent of the intertidal areas in China and Korea have been "reclaimed" or otherwise lost over the past three decades (Oldland et al. 2009:2–5). But in addition to growing human pressures on the landward side of many coastal areas, in some places sea-level rise may well be reducing this thin band of space from the seaward side, producing intensified patterns of "coastal squeeze" (Oldland et al. 2009:5).

Back in Manly, for all the difficulties faced by the Little Penguins, there are still grounds for hope. This colony has not yet gone the way of almost all the others on the Australian mainland. In this context, while the urban environment clearly poses a range of significant challenges for penguins, it is also important to note that it does offer some advantages, too. Living at close quarters with people brings houses and swimming pools, dogs, lights, and jet skis, but also vocal and organized advocates. Foremost among these people are the Penguin Wardens, a group of volunteers who spend their nights checking that no harm comes to those birds nesting under the busy ferry wharf. In addition, some members of the wider community have been actively involved in efforts to ensure that the National Parks and Wildlife Service does everything in its power to look after the penguins. In some cases, this has meant an active campaign of fox baiting to reduce the population of these (possible) predators. In one case, after several penguins where killed in a short period of time by dogs and foxes, it even involved the use of infrared cameras and the hiring of a sniper (van Dooren 2011a). And so the city also brings with it some advantages for struggling penguins (even if not for foxes). It is perhaps no coincidence that two of the remaining three mainland Australian colonies of Little Penguins are found in large cities: the Manly colony in Sydney and the St. Kilda colony in Melbourne. This fact gives me some hope for more sustainable human–penguin relationships in urban environments. But many of these places have already been lost, and those that remain are threatened. In short, there is much to do.

At this time in Earth's history, there are numerous different paths to extinction, numerous ways to end the possibility of life for generations of a kind. For many creatures, for many populations and species, it is the loss of an important place that leads inexorably to their end. In this context, perhaps the very least that we can do is begin to learn a new sensitivity to the storying and place-making practices of these nonhuman others, a sensitivity that just *might* provide an avenue to more sustaining possibilities of life, across species and generations.

Four

BREEDING CRANES

The Violent-Care of Captive Life

In December 2001, whooping cranes (*Grus americana*) migrated to Florida from Wisconsin behind a flimsy aircraft; these were the first whooping cranes to migrate in eastern North America in a century.

DAVID H. ELLIS ET AL., "MOTORIZED MIGRATIONS"

As we approached the enclosure, I could see several young birds moving around in the water on their long delicate legs. Standing about waist high and covered in the light brown plumage of their age, they looked very different from the much larger, mostly white, adult Whooping Cranes (*Grus americana*) that they would hopefully one day become. It was in the image of these adult birds that I was now dressed, wearing a long white costume with a hood and mask that almost completely obscured my human form. Joe Duff, my guide in this strange space of interaction, was dressed in the same way, but in one hand he carried a puppet—a lifelike model of a Whooping Crane head on a long neck that he used to interact with the birds, showing them how to peck and explore their surrounds in search of food.

Joe was the lead pilot for Operation Migration, the organization that was tasked with overseeing the care of these young birds. While they had been hatched and initially cared for at the U.S. Geological Survey's Patuxent Wildlife Research Center in Maryland, at between one and two months of age they were put on an airplane and flown here, to the White

A juvenile Whooping Crane with most of its adult feathers. (Ryan Hagerty/U.S. Fish and Wildlife Service)

River Marsh Wildlife Area in Wisconsin. This was to be their home for the next few months. During this time, Joe and the other staff from Operation Migration would continue to train them so that when the time came, they would be ready to take to the air and follow an ultralight aircraft on their first long migration south for the winter.

As Joe and I stepped slowly and carefully into the enclosure, the six young birds inside greeted us. As I moved past Joe to allow him to close the gate behind us, I remembered his advice to step slowly and avoid moving backward; he had explained that with the limited visibility afforded by the costumes, it was easy to step on a chick and cause real harm. The birds gathered around us, clearly paying close attention to our movements. Joe used his puppet to interact with several of them, while a couple of others moved over to a large shallow basin filled with water. In addition to wearing costumes to cover our physical appearance, we were not allowed to speak within earshot of the birds, so I slowly, silently, followed Joe as he moved around the enclosure.

For the most part, the birds followed Joe, too, exploring the environment and occasionally Joe himself—closely studying and even pecking at his costume as they went. They were clearly more interested in Joe than they were in me: perhaps this was because he was more actively engaging with them; perhaps they recognized him (his costume or his movements?) from frequent visits; or perhaps it was because I was not carrying a crane puppet, and so while I did not look like the humans whom these birds may have occasionally seen in the first weeks of their lives, I may also have looked distinctly "un-crane-like."

After just a few minutes in the summer sun, the costume was already stifling. I did not envy Joe or the other staff at facilities like Patuxent who spend long periods each day in the full sun dressed like this, taking birds for walks and swims to ensure healthy development. Between the limited visibility, the silence, and the heat, the experience quickly began to feel somewhat surreal. As I moved around, I was acutely aware of the presence of these small beings who occasionally looked closely at me, sometimes even following me. I wondered what they made of me in my costume: Did they think I was a crane or even a parent? How did these birds come to be in this place, entangled in such complex and intimate practices of daily care dedicated to both their own individual flourishing and that of their

Young Whooping Cranes in their enclosure, viewed through a peephole, at the White River Marsh Wildlife Area, Wisconsin. (Photograph by author)

species? What possibilities for crane life, and for human–crane relationships, emerge within this strange space of captivity?

This chapter is an attempt to answer these and other questions that occupied my mind that day as I spent time in the company of one of North America's most endangered birds. When I visited these cranes at White River Marsh in June 2012, they were already well into their intensive migration training. A couple of months later, they would leave Wisconsin on an almost 1,250-mile (2,000 km) journey south to Florida, with Joe Duff and others from Operation Migration leading the way. Having been absent from the skies and landscapes of the eastern United States for more than 100 years, Whooping Cranes have been reintroduced and retaught an ancient migratory route by dedicated humans (USFWS 2007:14).

The Whooping Crane story begins much earlier than this, though. Once present in populations across North America, by the early twentieth century the species was reduced to fewer than twenty birds. As a result of a range of factors, most importantly hunting and wetland loss, a single tiny population of Whooping Cranes was all that remained. These birds spent their summers breeding in Canada (in an area that straddles the border of Alberta and the Northwest Territories) and their winters on the Gulf Coast of Texas. The species was listed as threatened under federal legislation in 1967, before the passage of the Endangered Species Act (1973) (USFWS 2007:xi). Early conservation efforts focused on protecting summering and wintering grounds—now Wood Buffalo National Park, in Alberta, and Aransas National Wildlife Refuge, in Texas, respectively—while also attempting to prevent hunters from shooting birds on their long migration between these two areas, a struggle that continues to this day. These efforts have been slow but, over time, relatively successful: as of the spring of 2011 the Aransas–Wood Buffalo Population now numbers 285 birds (USFWS 2011:6068).

But conservationists long worried that while all Whooping Cranes remained within a single population, the species would continue to be highly vulnerable to extinction—as a result of infectious disease or a localized disaster (the greatest likelihood being a toxic spill along the shore of their wintering grounds in Texas, which has some of the heaviest barge traffic in the world, much of it for petrochemicals [USFWS 2011:6068]). As a result, from the 1960s the American and Canadian governments, along with many local partners, have run a captive breeding program, both to ensure the maintenance of valuable genetic diversity in captivity and to produce young birds who might be released in an effort to establish additional self-sustaining, free-living populations.

And so it is as part of this long conservation history that over the past decade, young birds have been trained each year to follow ultralight aircraft, as will be discussed in detail later in the chapter. This training is necessary because cranes raised in captivity will not migrate, not having been taught that they should—or when or how to—by their parents. To rectify this problem, ultralight aircraft were combined with an elaborate training program to establish the Eastern Migratory Population (EMP). Hailed by many as a great success, this population consists of birds who were taken south to Florida for the winter by Operation Migration, but who have usually returned to Wisconsin in the summer and then migrated south again the following year, both trips without any further direction or assistance (Duff, pers. comm.; Ellis et al. 2001, 2003).[1] This approach to teaching birds to migrate was pioneered by pilots Bill Lishman and Joe Duff, working with Canada Geese (*Branta canadensis*) and then Sandhill Cranes (*Grus canadensis*) in the 1990s.[2] Later, with the support of Patuxent and the International Crane Foundation, the technique was further refined with Sandhills and ultimately, in 2001, put to use with endangered "Whoopers" for the first time.[3]

This larger tale of Whooping Crane conservation—from a handful of birds to an increasingly healthy population alongside great efforts toward the establishment of additional populations (like the EMP)—is in many ways undeniably a story of intense and dedicated care. It is this care that has successfully held this species in the world, however tentatively, for four decades more than it might otherwise have survived. As a program guided by the imperatives of "conservation," care for the Whooping Crane *species* is central here. In a captive breeding program like this one, however, the goals of conservation are achieved through the dedicated daily care of the fleshy and willful *individual* cranes that comprise the captive population. Care for the species and care for individual birds come together here. As we will see, in various contexts the flourishing of individual lives and that of the species are bound together in mutually reinforcing ways.

In other ways, however, the conservation of the species has required that the good of individuals—Whooping Cranes and those of a number of other species—be "sacrificed." In order for the species to continue, captive birds have been required to live in strange and diminished environments and be exposed to ongoing stresses, including those of artificial insemination. Indeed, the simple fact of being raised in captivity has posed a range of developmental issues for Whooping Cranes, sometimes undermining their capacity to form social and reproductive relationships with other members of their species (for example, as a result of "cross-species imprinting"). Beyond the Whooping Cranes themselves, a number of other species have been drawn into this conservation project in a range of ways. These are the "sacrificial populations" that live their lives and die their deaths in the shadows, often unseen and unacknowledged, but making possible our hopes and dreams for the ongoing life of this species.

And so alongside intense care and hope, the recovery of the Whooping Crane is equally a story of violence and coercion. These are not mutually exclusive stories. Rather, in this time of extinctions, it seems that care and hope are frequently saturated with, perhaps even grounded in, unavoidable and ongoing practices of violence. All too often, this is how hope is produced, and care is manifested, at the dull edge of extinction. In fact, it is through these conservation practices that extinction has been further drawn out and "dulled" for many species, with a range of complex consequences—for good and ill—for all those involved.

This chapter is an exploration of the multifaceted and entangled regimes of care that animate and guide our interactions with Whooping Cranes in this space. As Maria Puig de la Bellacasa (2012) reminds us, "caring or being cared for is not necessarily rewarding and comforting." Caring is not achieved through abstract well-wishing, but is an embodied and often fraught, complex, and compromised practice. Simultaneously, it is "a vital affective state, an ethical obligation and a practical labour." In this context, caring is grounded in the mundane and "inescapable troubles of interdependent existences" and can offer no guarantee of a "smooth harmonious world" (Puig de la Bellacasa 2012:197–99).

Various caring practices come together here—sometimes in alignment and other times in conflict. Intimate care for some feathered bodies, some species, sits alongside the domination, coercion, and abandonment of others, giving rise to what I am calling "regimes of violent care." Delving into the processes and practices that underlie captive breeding efforts, this chapter explores how it is, materially and practically, that a species is held in the world in this way, and with what consequences for whom (Haraway 1997). Much of this book has (I hope) been an enticement to care for endangered birds. This chapter explores some of the forms this care might take. What does it mean to care for a species at the edge of extinction? What are the ethical stakes, the problems and possibilities, for interspecies relationships inside this contested space?

OUT OF THE WILD: THE EMERGENCE OF A CAPTIVE BREEDING PROGRAM

The extensive, long-term captive breeding program that lies at the heart of Whooping Crane conservation began in the 1960s. It was built on the simple practice of removing "extra" eggs from the Aransas–Wood Buffalo

Population. Most Whooping Crane pairs lay two eggs, but only one of them usually leads to a chick who successfully fledges. Conservationists reasoned that if—and this was a big "if"—they could remove these second eggs, successfully incubate and hatch them, and fledge the resulting chicks, this practice might lead to a substantial increase in the overall population (Ellis and Gee 2001).

For nearly thirty years, from 1967 until 1996, this is precisely what happened, enabling the establishment of a network of five captive breeding facilities across the United States and Canada (USFWS 2011). Of these facilities, the Patuxent Wildlife Research Center in Maryland (Patuxent) and the International Crane Foundation (ICF) in Wisconsin are now home to the vast majority of the world's captive Whooping Cranes. As noted earlier, the goal of these captive facilities is twofold. First, through the ongoing exchange of eggs, and occasionally live birds, they maintain a captive population that represents as much of the remaining genetic diversity of the species as possible. Second, these facilities rear young birds for release in an effort to establish and then maintain free-living flocks.

The assumption that many people have on first hearing about a captive breeding program like this one is that it is composed of a number of pens in which pairs of birds lay eggs and rear chicks. This approach to captive breeding is called "parent rearing." That it has a specific name at all is indicative of the fact that it is viewed as only one method of captive breeding among many. In the case of Whooping Cranes, it is an approach that has rarely been employed. From a conservationist's perspective, the key problem with parent rearing is that it does not produce enough chicks. Even given optimal conditions, an undisturbed pair of Whooping Cranes can raise only two young in a season. But if the same birds' eggs are taken away as they are laid, the birds can be encouraged to lay many more. John French (pers. comm.) sums up the situation simply:[4]

> Because these birds don't breed very readily in captivity, we want to maximize the output of any breeding female. So, we know that they lay a clutch of two eggs. What we're able to do is manipulate that clutch with dummy eggs to have them keep producing more eggs. For most pairs we get more than two eggs. Sometimes we get three, four, five, up to seven

> or eight eggs out of a single female. And we do that by removing the egg as its laid. If it's the first egg we replace that egg removed with a dummy egg—so that she stays in breeding condition—and then as second eggs are laid we remove them. The female then tries to complete the clutch of two. Those later eggs are typically a little smaller, a little lighter, and the chicks have a lower hatchability as the number of eggs laid goes up. But still, it works pretty well.

But these removed eggs have to be incubated, hatched, and reared by someone. And so, over the years, these tasks have often been delegated to birds of other species, to machines, or to humans: eggs have been variously incubated by Sandhill Cranes, domestic chickens, and/or machines; then hatched under Sandhill Cranes or by a mechanical hatcher; before finally being reared by Sandhill Crane or human surrogates, sometimes wearing a costume like that described earlier and sometimes not (Olsen, Taylor, and Gee 1997). In most cases, captive Whooping Cranes have not been involved at all in the rearing of their young, many being deemed to be inexperienced or poor parents who have in some cases failed to properly incubate or have even broken their own eggs.[5]

This high-volume approach to breeding was particularly important in the early years of the project, when the captive population was still being established. And now, since the collection of eggs from the Aransas–Wood Buffalo Population has been discontinued, these captive facilities have to produce all the young necessary to establish any potential new flocks. In an interview, Bryant Tarr estimated that if parent rearing were employed at ICF, the program might be able to raise three or four chicks in a year, and Patuxent perhaps six to eight. Instead, with the help of a range of human and nonhuman surrogates, they are able to rear dozens of birds each year.

But involving surrogates in the early lives of Whooping Cranes also raises many potential problems. Perhaps most important for birds who one day will be released, close contact with humans (or, indeed, with surrogates of other species) risks young Whooping Cranes becoming habituated to, or "imprinted" on, individuals of another species. These relationships are potentially deeply problematic for both the imprinted birds and the continuity of their endangered species.

REPRODUCTION AND MIGRATION: THE PROBLEMS AND PROMISES OF IMPRINTING

The study of imprinting is intimately tied to the life and work of the noted Austrian zoologist Konrad Lorenz (1903–1989). Through his experiments with a range of birds—perhaps, most memorably, with Graylag Geese (*Anser anser*) and Jackdaws (*Corvus monedula*)—Lorenz (1937, [1949] 2002) played a key role in defining and popularizing the phenomenon of imprinting (Hess 1958). He showed that many species of birds do not instinctively recognize their parents, or indeed others of their own species, but are "conditioned" into this knowledge in the early stages of life (Lorenz 1937:262–63). Lorenz identified a small window of time in the first few days after hatching during which chicks form a deep attachment to the first moving object that they encounter. Ordinarily, this moving object is a parent, and imprinting in this way enables birds to keep their young close, especially those with "precocial" young who are able to move around very quickly after hatching and might easily be lost (Hess 1964:1132).

In his experiments, Lorenz (1937:264) took eggs and newly hatched chicks and exposed them to a range of alternative "parents": ducks hatched and raised by geese; geese raised by turkeys. In one experiment, a parakeet was raised in isolation to determine whether he would imprint on a ball hanging by a string in his cage; he did (Lorenz 1937:270). In several other cases, Lorenz himself was the object of these imprintings, giving rise to the now iconic photos of Lorenz as "goose mother": a man walking in a field attentively trailed by a line of Graylag Geese chicks.[6]

But Lorenz (1937:263) quickly realized that imprinting plays a far more significant role in many birds' lives than simply determining who or what they will follow as hatchlings. Although the details vary between species as well as between individuals within a species, for many birds, including Whooping Cranes, imprinting plays a profound role in establishing a bird's understanding of its broad social group. As a result of this developmental process, birds that are reared by "parents" of another species—whether another bird species or humans—tend to preferentially associate with individuals of these kinds, individuals that look and sound like their "parents." This tendency continues throughout life, and so when it comes time to choose a mate, a bird that has been imprinted in this way usually attempts a cross-"species" relationship. Lorenz's (1937) parakeet offers a good, if somewhat disturbing, example through his efforts to court the ball in his cage: "[H]e would execute exactly the same movements [as any other male parakeet courting a female], but, as he was aiming them in such a way that the ball represented the female's head, his thrust-out claw would grip only vacancy below the celluloid sphere dangling from the ceiling of his cage" (270).

Today, ornithological studies suggest two separate, but related, periods of imprinting in cranes' and some other birds' early development. The first occurs in the early days of life, as Lorenz suggested (often called "filial imprinting"). A little later—in Whooping Cranes, probably around the age of ten to fourteen weeks, just before fledging—it is now thought by many, that a second period of imprinting occurs (often referred to as "sexual imprinting" [Swengel et al. 1996:119]). It is at this latter point—or through some combination of experiences at both—that young cranes form their understanding of an appropriate mate based on the kinds of beings that they are associating with (usually their parents and possibly also members of a larger flock) (Horwich, Wood, and Anderson 1988; Swengel et al. 1996:119). It also seems that there is a great deal of variability among (even closely related) species in the extent to which these early experiences shape birds' mate selection later in life (Slagsvold et al. 2002). As we will see, for Whooping Cranes—as for many, perhaps most, other bird species (Immelmann 1972; ten Cate and Vos 1999:6)—this kind of imprinting plays a central role in the selection of a partner.

In 1982, many North Americans got their first glimpse of this avian potential for strange couplings, courtesy of the Whooping Crane project. George Archibald, a renowned crane biologist and then director of ICF, appeared on *The Tonight Show* to tell the story of his relationship with a Whooping Crane named Tex (Butvill 2004; Hughes 2008:143). Tex was the only surviving offspring of a pair of Whooping Cranes at the San Antonio Zoo. As such, her genetic material was deemed to be extremely valuable for the survival of the species. However, despite years of breeding attempts, she had not successfully produced any young. So in 1976, at the age of ten, she was transferred to ICF. But there, too, she showed little interest in other cranes.

It seemed that Tex, who had spent the first weeks of her life in the home of the director of the San Antonio Zoo, had been imprinted on a human. At ICF, she developed an attraction to Archibald. Tex could be artificially inseminated, but "she would not produce a fertile egg unless her hormones were jump-started by courtship. Archibald would have to win her heart" (Hughes 2008:142). To this end, he set up an office in Tex's indoor pen and spent his days with her. He walked and danced and sang with her—doing his best impersonation of a Whooping Crane. As the breeding season approached each year, they moved out to a larger field to establish a territory and together build a nest. Archibald spend his days in the field with Tex, working at a desk in a tiny shed. For seven years, he and Tex maintained this relationship, producing many eggs but none that hatched successfully. When Archibald was overseas, Tex maintained her monogamous relationship with him, refusing to keep company with anyone else (Archibald was married throughout this period). Then, finally, in 1982 they successfully hatched and fledged a chick, a male who was named Gee Whiz. Despite a rocky start to life, Gee Whiz survived and went on to produce many offspring, ensuring the continuity of his genes. Archibald and Tex did not produce any more young; about three weeks after Gee Whiz hatched, a raccoon got into Tex's pen and killed her (Butvill 2004).

This sad story serves as an example of the powerful sexual connection that is established through imprinting, a connection that can be particularly problematic when humans and other species take on surrogate rearing roles, becoming entangled in the reproductive lives of birds. In the early years of the Whooping Crane project, many birds were raised in such a way that they developed cross-species social and sexual cues, as had Tex.

It was precisely this kind of cross-species imprinting that is thought to have been behind the failure of one of the first efforts to establish a new free-living population of Whooping Cranes, at Grays Lake National Wildlife Refuge in Idaho. In this case, instead of producing chicks in captivity for release, conservationists tried a different approach to maximizing the number of young cranes. Beginning in 1975, Whooping Crane eggs were placed directly under free-living Greater Sandhill Cranes (*Grus canadensis tabida*) in place of their own. These eggs were collected from both the Aransas–Wood Buffalo Population and captive birds at Patuxent. This approach, called "cross-fostering," left all subsequent incubating and rearing up to the adoptive Sandhill parents.

Sure enough, when cross-fostered Whooping Crane eggs hatched, the adoptive parents diligently taught the chicks to find food and migrate, but at the same time imprinted on them a social identity that would ultimately lead to lives of isolation and exclusion. According to biologist Janice Hughes's (2008) careful piecing together of events at this time: "Chick-parent behavior had been typical through early winter, although foster chicks were frequently harassed by other adult Sandhill Cranes. Also, families with foster chicks were never really accepted into sandhill society and tended to avoid confrontation by loitering on the periphery of winter-feeding flocks" (164). This kind of liminal existence, on the edge of the Sandhill flocks that they were hatched into, came to characterize the lives of these Whooping Cranes. Once they left their familial units—many doing so well before Whooping or Sandhill Cranes normally would—they "remained ostracized wherever they traveled" (Hughes 2008:164). As a result of this positioning, these birds were often placed at greater risk of starvation, predation, or being shot by hunters.

The greatest difficulty, however, arose when these cross-fostered birds reached sexual maturity. It seems that they had little interest in other Whooping Cranes. Even when a captive female was deliberately released into the territory of a cross-fostered male, he displayed generally heightened aggression to Sandhill males, but seemed uninterested in (and perhaps confused by) the Whooping Crane female—although even if he had been interested, it has been speculated that these cross-fostered cranes may well not have known the appropriate Whooping Crane calls and dances to effectively express this interest (Hughes 2008:165). Other solitary Whooping Crane males built nests, "one insisting on incubating an empty nest. Two other males assisted a sandhill pair with raising their chick—not typical behavior for a highly territorial, long-lived monogamous species" (Hughes 2008:167). In short, these cross-fostered cranes displayed a great deal of unusual reproductive behavior.

In all, 289 Whooping Crane eggs were cross-fostered in this way over roughly a decade before the program was deemed to be a failure and cancelled in 1983 (Ellis and Gee 2001:17; Hughes 2008:167). From these eggs, a total of 77 Whooping Crane chicks fledged and migrated, but there is no evidence that any of them ever formed pairs and successfully bred—with the exception of a single cross-species pair that produced a hybrid chick named Whoophill by researchers (Hughes 2008:166).

Imprinting clearly raises problems for a cross-fostering approach to producing new Whooping Crane populations, but rearing cranes in captivity can be equally problematic, often leading to cranes imprinting on humans instead. As we have seen, parent rearing is often not considered a viable option because of the small number of chicks that are produced. As such, human keepers are required to be intimately involved in the daily lives of young birds, teaching them to look for food and encouraging the walking and swimming necessary for healthy development. But humans must somehow be present without becoming the object of imprinting. They must not only cover over their own human form to *prevent* birds from imprinting on it, but at the same time produce the form of an adult Whooping Crane, so that birds *can* imprint on *it*.

The costume that I wore at White River Marsh is, of course, a big part of this effort, but only a part. The generally accepted approach at both Patuxent and ICF is for eggs to be incubated by a combination of Sandhill Cranes (for roughly the first two weeks) and machine incubators, before being hatched in a mechanical hatcher and reared by human staff. To prevent chicks from imprinting on humans, costumes are combined with an elaborate system of screening and the introduction of live and artificial "imprint model" adult birds—who are kept visible and audible to young cranes. The actual day-to-day rearing work is done by silent staff wearing masks and shrouds and using crane puppets to interact with the birds, while also carrying small audio devices that play comforting parental vocalizations (Duff et al. 2001; Wellington et al. 1996).

Even after young birds destined for release have passed through this delicate developmental stage and are no longer at risk of imprinting on their human keepers, they still need to be costume reared. Otherwise, frequent exposure to humans—especially helpful humans who provide care and food—would risk habituating them. In the early years of the Whooping Crane project, while experiments were being conducted to develop costume protocols and other captive breeding and release techniques (often on Sandhill Cranes), many birds were thought to be overly habituated to humans when they were released. These cranes visited schools and suburban areas; three experimental Sandhills even took to spending their days in a state penitentiary in Arizona (Duff et al. 2001:115; Ellis et al. 2003:262). This behavior caused concern on two fronts: that such big birds might harm people, or that the birds themselves might be harmed by failing to avoid dangerous urban environments. And so costume-rearing protocols are now utilized for all birds that will be released into free-living populations, for the entirety of their time in captivity.[7]

However, alongside the many *problems* that imprinting has raised for the captive rearing of Whooping Cranes, it has also presented some powerful *possibilities* for the production of human–crane relationships that might aid in the continuity of the species. In particular, it is these birds' ability to accept a costume-wearing human as a parent and then diligently follow that parent wherever he or she goes that has enabled the development of the ultralight aircraft–led assisted migration. As the "when and where" of migration are taught to free-living young birds by their flock (usually their parents), captive-reared and released birds have to be taught to migrate. And so, in addition to enabling more young birds to be hatched each year, costume rearing has been preferred to parent rearing because it holds the promise of the creation of new *migratory* populations of Whooping Cranes.

In order to have cranes follow planes, however, they have to be introduced to the aircraft from a very early age. In fact, even while still in the

A costumed carer seated in an ultralight aircraft interacting with young Whooping Cranes. This feeding is part of the training that familiarizes cranes with the aircraft, which will be used to lead them on migration. (Paul K. Cascio/U.S. Geological Survey)

egg, birds that will be taught to migrate in this way are exposed to the sounds of the ultralight's engine and propeller in the incubator. Approximately a week after hatching, chicks are shown the aircraft itself. This introduction takes a somewhat peculiar form: it involves a costumed person sitting in the aircraft, with the engine running, feeding cranes with a long puppet head extended from the cockpit. In this way, the potentially frightening sight and sound of the airplane are coupled with the reassuring presence of a "parent" and tasty mealworms. When the chicks are just a little older, the wing of the aircraft is removed to make it more maneuverable on the ground, and the chicks are encouraged by a costumed handler to follow it—first within a relatively small enclosure, and then on longer walks in a field. Similar training continues, on the ground and then in the air, reinforcing the bond between the chicks and the costumed pilot. Eventually, after months of preparation, the young birds are ready to fledge, and together they set off behind the plane on their first long journey south (Duff et al. 2001).[8]

THE ETHICS OF IMPRINTING: COERCION AND CAPTIVITY

While imprinting clearly raises a range of practical problems and possibilities for the captive breeding of cranes, the ethical dimensions of the specific kinds of relationship that are possible through imprinting have received little attention. In an ethical context, the first thing to note about imprinting is that it occurs at an early developmental stage in a bird's life—when a chick is open to the world in a specific, deeply consequential, way. As we have seen, imprinting can have a profound impact on a bird's understanding of its social and reproductive world. In particular, as Konrad Lorenz (1937) noted, imprinting is characterized by a strange fixedness (Hess 1964). Once an object-attachment has been formed through imprinting—whether to a biological parent, a human, or a ball—this constitution of a bird's social world is very difficult, in some cases perhaps impossible, to change (ten Cate and Vos 1999). In an important sense, then, imprinting is not like many other human–bird interactions; it is not about the formation of a relationship between two subjects, who—however unequally positioned—already have a significantly well-formed way of life, a way of being in the world produced through particular biosocial inheritances and individual experiences (chap. 3). Rather, imprinting enters into and disrupts some of these modes of inheritance, taking advantage of an ontological openness to produce an altered way of life.

Of course, all subjects are formed in and through their interactions with others. Michel Foucault (1980, [1975] 1995) taught us about the always unequal dynamics of subjectification, in which living beings are formed and re-formed through their entanglements with technologies, discourses, and institutions. In a multispecies context, Donna Haraway's (2003) work has charted some of the many ways in which organisms are enfolded in webs of intra-active becoming: "Beings do not pre-exist their relatings" (16). Anna Tsing (2012) reminds us that this goes for humans, too, that what we think of as "human nature" is, over both evolutionary and personal time frames, "an interspecies relationship" (141). And so my comments about this entry of people into birds' lives are not premised on a refusal to accept the co-shaping of all subjects in and with members of other species, but on a recognition of the *particular* modes of being and becoming with others that occur in imprinting.

In her fascinating and provocative discussion of Lorenz's relationships with Jackdaws and Graylag Geese, Vinciane Despret highlights the co-becoming of human and bird in relationships of imprinting. As she notes, these are not simple cases of bird becoming human, or human becoming bird. Rather, they are examples of an "anthropo-zoo-genetic practice" that constructs *animal and human* (Despret 2004a:122): "Lorenz and his goose, in a relation of taming, in a relation that changes both identities, have domesticated one another" (130). Despret is particularly interested in the way in which Lorenz's approach to living with his experimental subjects enabled the formation of new kinds of relationships—embodied, caring relationships—that hold open possibilities for new kinds of knowledge production in which all the participants, as well as the research questions themselves, are at stake, open to being rearticulated.

I find Despret's work to be provocative and profoundly helpful in rethinking possibilities for human–animal research. And yet, there is something about the dynamics of imprinting, so central to Lorenz's work, that I think complicates the ethics of these relationships. In simple terms, imprinting is not taming; these are fundamentally different developmental processes in a bird's life. Both can be, and often are, coercive; but they are coercive in different ways. In particular, the space for agency and resistance is not at all the same. At points in her discussion, Despret seems to follow Lorenz in presenting his formation of new relationships with young birds as though they were initiated by the birds, or at least jointly initiated. For example, when discussing Marina, a goose that Lorenz kept and observed for the few hours after her hatching and then tried to return to another bird's care, Despret (2004a) notes that the young goose "refused to stay, and addressed to Lorenz a desperate 'abandoned call' ": "Lorenz tried, but could not convince her not to follow him. Then, he says, I behaved exactly as if I had adopted her, pretending to ignore that, in fact, it was she who had adopted me. For the whole day, and the coming days and months, Lorenz played the role of a good goose mother" (129). The agency is all mixed up in this account. Given what he *already knew* about imprinting, Lorenz's removal of Marina in the first hours of her life *was* an act of adoption. Her refusal to stay with other birds afterward was utterly predictable. As such, it is not helpful to present the situation as one in which Lorenz responded to the young goose's demands, as though she had adopted him. All of Lorenz's relationships with imprinted birds have to be read in this same light. For good or ill, irrespective of the valuable scientific knowledge that they produced—knowledge that now contributes to Whooping Crane conservation—these were fundamentally coercive relationships, in which one partner knowingly manipulated the delicate developmental stages of the other to produce a lifelong attachment: a *captive* form of life.

This leads us to another important difference between imprinting and many other kinds of relationship: the consequences of imprinting, for a bird's sense of self and world, are profound. Imprinting does not produce a relationship in which human and bird enter into each other's social worlds, leading to additional possibilities for connection and care. Rather, it produces a relationship with humans *at the expense of* a whole set of other ways of being, often severing the possibility for a bird's relating with others of its own species, and so profoundly altering its chances for social and procreative relations. All too often, birds are pulled into this space of complex interspecies-subjectivity through human carelessness: imprinted on their keepers or another species of bird, and then abandoned ("released") into a lonely world in which their social and sexual cues are out of kilter with those of their conspecifics. I cannot help but think here of the countless Whoopers, imprinted on Sandhills within the Grays Lake

While Lorenz ([1949] 2002) clearly cared for his birds in many ways—diligently taking on the role of parent or mate: hourly feedings for some young birds and himself being fed worms by a Jackdaw suitor—all this care cannot be extracted from the broader framework of coercion, captivity, and violence within which it occurred. While I agree with Despret (2004a) that Lorenz's experimental approach was grounded in relationships of care that enabled the formation of new kinds of knowledge, my concern is that his technique, while good for learning, may not have been so good for geese. The sheer number of birds that Lorenz reared in this way, as well as the seemingly carefree way in which he mixed the birds' social and reproductive worlds (including with inanimate objects), leaves me with little doubt that he saw no real ethical problems with his experiments. This broader framework of coercion makes me inclined to view Lorenz's subjects in much the same light that Douglas A. Spalding saw the chicks in his own very early experiments on imprinting, not as birds deeply cared for but as "little victims of human curiosity" (Sluckin 1964:2).[10]

And so I am left with the view that insofar as an ethical relationship with a deliberately human-imprinted bird is possible, it requires a genuine commitment grounded in ongoing and dedicated care for that individual being. I am highly dubious about whether the affective and ethical obligations, as well as the real time and daily labor of being with and providing for another, can be adequately achieved if the individual bird and its well-being are not the primary motivation for the relationship. When care for the individual comes a constant second to the generation of novel scientific knowledge or the conservation of the species, it is just too easy for the kind of vulnerable and captive lives produced through imprinting to be neglected.[11]

In the context of Whooping Crane conservation, this discussion of the ethics of cross-species imprinting is important on two fronts. First, although not currently a part of the whopping crane program, the deliberate imprinting of birds on humans is a central component of many avian captive breeding efforts. Understanding the nature of the relationship produced through imprinting—as best we can—is central to evaluating the ethics of the use (or non-use) of this practice (as discussed later in the chapter). Second, this critical understanding of imprinting, through contrast, helps us to better appreciate the contemporary efforts of Whooping Crane people to rear their charges. However messy, flawed, and imperfect costume rearing is, it is a clear groping toward a more ethical form of life for people who have tangled themselves up in the social and reproductive lives of birds. In addition to concealing the human form (to some extent), costume rearing enables humans to participate *with care* in the social and developmental worlds of young cranes.

This is not about humans "becoming" cranes in any full sense. The staff members whom I spoke to at Patuxent, ICF, and Operation Migration had different ideas about the extent to which a person needed to walk and move like a crane when in costume, but everybody acknowledged that costumed humans can do only a very imperfect job. There is no perfect assumption of a crane's way of being, nor can there be an absolute covering over of the human presence. Instead, in Despret's (2004a:122) terms, this remains an "anthropo-zoo-genetic relation" in which both parties will be unavoidably changed by their association. And yet, costume rearing is an attempt to *distribute* the impacts of this relationship in such a way that humans take on more and more of the burden of comporting themselves like another, so that cranes with a better chance of flourishing lives might emerge.

As noted, this process of comportment is far from perfect. It is an ongoing effort. Decades of studies at Patuxent, ICF, and elsewhere have gone into better understanding what it is that young Whoopers imprint on: Is it certain colors, shapes, sounds, or a combination of all these? Long red objects, like the cap on an adult crane's head, are thought to be particularly important, as are comforting vocalizations, the physical act of following a parent, and the presence of live and dummy "imprint models" (Swengel et al. 1996; Wellington et al. 1996). Ultimately, staff can only hope that through this elaborate process, young birds are imprinting on something like a Whooping Crane—not on people and certainly not on ultralight aircraft. The fact that many of these birds go on to form monogamous relationships with other Whooping Cranes is a very good sign. Alongside this research, rearing cranes with care requires the daily labor of numerous keepers. Young cranes need an awful lot of exercise for healthy development. For example, when raised in captivity, they have often developed toe and leg problems, presumably as a result of their not getting the amount of exercise that they would ordinarily get while attempting to follow their parents as they make huge strides through dense wetlands (Wellington et al. 1996). As a result, staff now spend hours each day in stifling costumes exercising their young charges on long walks and swims.

In this context, costume rearing might be understood as a practice grounded in what Traci Warkentin (2010) has called "interspecies etiquette." Acknowledging that we cannot ever have perfect access to the worlds of others, Warkentin's approach emphasizes the need for ongoing practices of embodied and caring attentiveness. This attentiveness, she argues, might provide the level of understanding necessary to live alongside, or even in intimate entanglement with, others in a way that enables the best chance for a flourishing life for everyone. Through both research and personal effort, costume rearing Whooping Cranes is an attempt to have humans be deeply implicated in birds' lives, while avoiding their imprinting on people. In so doing, it is an effort to undo some of the violence of captivity and ensure that possibilities for intra-species life are not foreclosed.[12]

But for all its merits, costume rearing is not motivated by an ethical agenda. The dedicated care for the development of individual cranes that is practiced at Patuxent and ICF is ultimately grounded in another, more fundamental, regime of care that drives this conservation project: the care for the species. As in so many other aspects of conservation work, costume rearing is guided by what Matthew Chrulew (2011a) has called "species-thinking," in which "each individual is only perceived as a token of its inexhaustible type" (141).[13] While a species and its individual participants cannot be disarticulated in any straightforward way—the "survival" of either requires the continuity of the other—each can still be cared for in ways that give more or less regard to the well-being of the other (at least in the short term).

As staff at Patuxent, ICF, and Operation Migration commented to me in interviews, costume rearing is driven by "functional" conservation outcomes; it is the approach that produces the maximum number of birds best adapted to survival and reproduction once released. If elaborate and costly costume rearing was not necessary for the survival of the species, it is highly unlikely that it would have been adopted for the well-being of individual birds. In fact, as will be discussed further below, in other captive breeding programs birds are commonly deliberately imprinted on humans to enhance their reproductive output.

There is an important tension here between what are often thought of as "conservation" and "animal welfare" agendas. I resist this framing of this situation, which tends to present conservationists as cold and unethical, while simultaneously creating a veneer of objectivity around their particular goals, which are presented as being those of "conservation science," not personal values. Instead, I understand this situation as a site of overlapping regimes of care. In this context, conservationists may more readily be understood as pursuing a different caring project, no less value driven than that of welfare activists. As Joanna Latimer and Maria Puig de la Bellacasa (2013) note, what a scientific community cares for—"how some concerns get made present and others made absent"—is an important aspect of the defining, shaping, and even legitimating of that field. In the context of any given scientific practice, which is always also a social practice, particular ways of caring, particular objects of care, are preferenced over others (albeit in ways that are not at all static or fixed). In the context of conservation biology and its material practices, it is clear that care for the species often trumps other considerations, including the well-being of individual animals. Sometimes, as in the case of costume rearing, a practice is able to contribute positively to the flourishing of the species and individual cranes' lives. In other cases, as we will see, these regimes of care fail to align and one agenda comes to dominate the other.[14]

SURROGATES, SACRIFICES, AND SPECIES-THINKING

The picture that I have sketched thus far is really of only one side of the world of Whooping Crane captive breeding: that of the young birds raised to be released. But, as in similar programs for other endangered species, the five captive breeding facilities in North America are populated by Whooping Cranes who are required to spend the entirety of their lives in small enclosures: breeding birds producing offspring to populate a world that they will never see. Leading largely hidden lives, these birds carry out the "invisible work" (Star and Strauss 1999) that animates our hopes for the survival of the species. A sharp divide exists here between the management and the lives of two relatively distinct populations, each closely managed to produce optimal subjects for very different ways of life. But Whoopers are not the only individuals drawn into this space of perpetual captivity: a host of other birds have been required to take up roles in this project as incubators, tasters, and testers of various kinds. These functional, surrogate, and often sacrificial populations—which are all in different ways integral to the possibility of ongoing Whooping Crane life—are the focus of this section.

The total captive breeding population of Whooping Cranes in North America is roughly 150 birds, primarily located at Patuxent and ICF. As is perhaps already clear, the reproductive lives of these birds are highly managed. Both facilities employ similar methods of ensuring that birds produce as many eggs as possible each season, removing eggs as they are laid in the way described earlier. In order to get birds to produce fertile eggs, however, artificial insemination (AI) is widely utilized. In some cases, AI is required: perhaps because the male is infertile or one of the birds is not adequately fertile, or clipped wings have undermined the flapping and balance necessary for proper copulation (French, pers. comm.; Gee and Mirande 1996). In other cases, a little "assist" is provided by means of AI, to ensure that fertilization has taken place. And in yet other cases, particularly at ICF, AI is utilized to create a genetic pairing that would not otherwise be possible, perhaps because particular birds have other partners or do not get along with each other. Highly managed genetic pairings produced through AI are increasingly common in the captive breeding of endangered species, since limited genetic diversity must be preserved in ways that do not always map well onto the actual fleshy, willful beings that "contain" it. For one reason or another, however, almost all the breeding cranes at ICF and Patuxent are subject to ongoing AI.

Artificial insemination is a set of material practices and technologies that is often paid very little attention outside the relevant scientific specialist groups. First successfully utilized in birds over a century ago, these techniques are widely employed in the commercial poultry industry (which continues to be the site of the production of much of our knowledge of avian AI), as well as in the captive breeding of more than forty species of endangered birds (Blanco et al. 2009:200). For all its conservation and commercial possibilities, however, AI remains a highly invasive and often stressful practice.

Several methods of AI are utilized in avian captive breeding programs. With Whooping Cranes, the predominant approach is called "abdominal massage." Here, a male bird must be captured and restrained by a small group of people, who then massage it to "encourage compliance" in the collection of semen:

> The male is cradled between an assistant's legs, with the bird's head and neck behind the handler [between his or her legs] and the animal's breast propped against the assistant's thigh. Safety for the bird and operator is critical. During the breeding season, cranes can be aggressive, and a broom is often used as an object for the male to attack and to help corral the bird for restraint. (Blanco et al. 2009:203)

Other approaches to restraining birds for abdominal massage have been employed for different species. Parrots are often placed headfirst into a clear plastic tube to prevent biting. The bird's lower half is still accessible for stroking, and its feet can be restrained to prevent scratching. Once semen is collected, birds are "easily released through the tube front" (Blanco et al. 2009:203).

As these examples indicate, abdominal massage is often highly stressful for the birds; many strongly resist and attack handlers. This is also not an infrequent activity, with some birds being "collected from" two to three times a week (Gee and Mirande 1996:207). The AI process for females is similar in many respects, with birds needing to be captured, restrained, massaged, and finally inseminated. The process is described in detail in a handbook of crane biology and husbandry, a section of which reads as follows:

> At Patuxent, female cranes are massaged just as for the males. In the ICF method, the female's back and sides (posterior to the wings) are stroked to simulate the male's abdomen on the female's back during copulation. Semen can then be deposited into the cloaca or vagina. . . . It is even possible to deposit semen in the oviduct of uncooperative cranes . . . the distal end of the oviduct can often be everted by placing firm pressure on the female's abdomen and the walls of the cloaca. We do not recommend that an inexperienced person evert the cloaca because force can cause injury and undue stress to the bird. (Gee and Mirande 1996:208)

These practices of artificial insemination are in stark contrast to the lives of those young birds who are costume reared and eventually released.

Once a crane enters the breeding population, it is usually handled out of costume. Many of these birds were costume reared initially—to avoid imprinting on humans—but once it has been determined that they will not be released, habituation to human presence is no longer a serious problem to be avoided. In fact, it is actively encouraged. For these birds, a certain amount of habituation is regarded as being not only inevitable, but necessary and desirable for the success of the program.

In addition to the obvious behavioral signs, intensive studies for the commercial poultry industry have clearly indicated that AI can cause high levels of stress for birds. It is widely thought that this stress may have a significant negative impact on successful avian reproduction (Blanco et al. 2009:202). As such, Whooping Cranes and other captive-bred endangered birds are now often "trained to accept the technique," encouraging familiarity with the process and the team involved (Gee and Mirande 1996:209): "At ICF, 12 species of cranes have been successfully inseminated after they assumed copulation posture and everted their oviducts in response to massage stimulation and handling" (208). The degree to which cranes come to accept AI varies among individuals, but with all birds some attempt is made to habituate them to the process in order to reduce their stress. Doing so increases the quality of the semen that is collected (reducing contamination by urine and feces) and improves the chances of successful fertilization (for a number of reasons), and it decreases the chance that staff or the birds themselves will be harmed during the process.

In other avian captive breeding programs, the potential to produce stress-free participants for AI has led to the development of a "cooperative" approach in which birds are deliberately imprinted on a human handler. As should be clear from the earlier discussion of imprinting, I would question the appropriateness of the term "cooperative" in this context. This practice is particularly popular in the captive breeding of birds of prey. As Tom Cade (1988) notes, these birds are frequently "raised in isolation to develop human-imprinted raptors into highly effective semen donors for artificial insemination. The birds are usually trained to copulate and ejaculate on a special hat worn by the human companion. Such birds produce copious volumes of high-quality semen for 2 to 3 months and have been great assets in increasing the percentage of fertile eggs laid in raptor breeding projects" (281).

This kind of deliberate imprinting of captive-bred birds on people is not practiced with Whooping Cranes at either Patuxent or ICF. If birds were to be imprinted on people, then breeding would require the kind of elaborate interactions described in the case of Tex and George Archibald. In an interview, Bryant Tarr (pers. comm.) explained that while imprinting on humans would offer significant advantages in small captive populations, with "a big flock of birds, we need them to 'work' each other, basically." There simply are not enough staff members to provide the significant hours of contact that a crane needs with its partner, especially during the breeding season. As such, it is deemed to be far simpler and more efficient to have cranes pair with, and play these courtship roles for, each other (and then use AI to manage genetic pairings).

It is difficult to know what to make of these practices of habituation and imprinting to facilitate AI for birds who will remain in captivity for the entirety of their lives. While these various approaches to making AI more "cooperative" are clearly an intrusion into cranes lives and perhaps developmental processes, they also likely decrease the stress of the captive environment and so perhaps improve these birds' quality of life. As Tarr (pers. comm.) suggested, it is helpful to understand captive birds' response to AI along a spectrum:

> I think that to a certain degree there's a gradient, from birds who are clearly distressed by it and not wanting to cooperate, and then there's a whole gradient of birds who are semi-cooperative, up to birds that love it. Imprints would be over here in the "love it" category. Unhandled wild birds would be on the other extreme. Our captive birds are somewhere in the middle, and that's on purpose. They really don't want us to come in and grab them and manipulate them, but when we do they just slowly walk away from us—they're not scared, they're just avoiding us—then we gently corner them in the corner of their pen and grab them and we put the bird's head between our legs [as described earlier]. Then we massage their back and thighs. At that point, a bird that is acclimated to the process sort of turns the curve from "I want to get away from you," to "hey, this is feeling okay"—and they start purring and relaxing.

And so this "middle ground" in the spectrum is deemed to be the most suitable and practical for Whooping Cranes in the captive breeding population. In interviews with staff at Patuxent and ICF, it was clear that, as in the case of costume rearing, this approach is driven by a range of largely functional factors geared toward maximizing conservation outcomes: cranes imprinted on their own species but habituated to people are more likely to produce a larger number of fertile eggs, are less likely to harm themselves or their keepers during AI and other interactions, and can pair with other cranes so do not require too much human participation in the form of dancing, singing, and other courting activities.

I suspect that this middle ground may often also offer the best ethical options: reducing stress from human contact while respecting some level of autonomy in the birds' daily and developmental lives and ensuring that they have a partner reliably present. Ultimately, however, it is impossible to say what the best option might be. This kind of ethical inquiry pushes the limits of what we can sensibly talk about. Ethics needs ethology here: case-specific research (which is itself often invasive) on the pros and cons of these various ways of life for those who are forced to live them.

For example, while deliberate cross-species imprinting may be a coercive act in general terms, can this notion be applied universally? In the case of those birds who are required to spend the entirety of their lives in a captive breeding population, it seems that being imprinted on a person so that they not only accept, but enjoy, the process of AI may actually be a route to a less stressful, and so perhaps more flourishing, life. In entering this fraught space, we are immediately confronted with the lurking "taboo of interspecies pleasure." I am not at all convinced that the simple fact of sexual pleasure across species boundaries is a negative that should be understood as undermining the desirability of this approach to captive life. If anything, the opposite seems to be true. In this specific case, however, this pleasure should not be separated from the deliberate intervention in developmental processes that has given rise to its expression: birds conditioned to "love their cage."

Again, however, the ethics of this relationship are unclear. While the prospect of conditioning other humans to accept treatment that they would otherwise find stressful and invasive, especially treatment of a sexual nature, is deeply disturbing, is it right to apply the same general principles to cranes? Are "free will" and "autonomy" really significant concerns here, or is translating them into this context itself a form of violence to captive birds? While deliberately imprinting cranes on humans or other species is a coercive and violent act, in some cases perhaps it is preferable to a life of fear and stress in captivity. Or perhaps habituation to AI and other aspects of captivity adequately overcomes these issues, so the ICF and Patuxent approach is the best that we can hope for. I am genuinely unsure about what the best options are in this difficult space, with its multiple and overlapping forms of "captivity."

But it is also important to remember that the stress of living with humans is far from being the only ethical issue that is raised by a life of captivity. No matter how well conditioned cranes are to humans, these birds necessarily lead diminished lives on a variety of other fronts. Spending most of their time in large grassy pens, their days do not contain anything like the environmental, social, or behavioral diversity experienced by free-living Whooping Cranes: not able to fly or to stride through wetlands in search of food, and most of them ultimately never given the opportunity to hatch their own eggs or rear their own young.

While it should not be forgotten that free-living birds also experience a range of deprivations and stresses—the world beyond the fence is not Eden—this does not change the fact that life as a captive breeder remains a kind of "sacrificial life," a life given, and not by one's own choice, for the good of others. It is a life sacrificed as part of an effort to care for a species. But it is also not only individual Whooping Cranes whose lives are drawn into the fray. In a range of ways, a host of individuals of other species are also positioned as sacrificial surrogates for the continuity of the Whooping Crane. Sandhill Cranes have tended to bear a large portion of this burden. At Patuxent, ICF, and elsewhere, captive colonies of these birds spend their lives incubating Whooping Crane eggs. While much of this incubation is now done by machine, it seems that nothing we are able to produce quite matches the skill of experienced parent cranes. The longer eggs can be left under Sandhills before being moved to the machine incubator (especially in their early stages), the better their chances of successful hatching. And so hundreds of these bird spend their lives behind fences for the roughly two weeks of skilled incubation that they can provide for each new Whooping Crane egg.

Similarly, when it came to establishing the feasibility and best practices for ultralight aircraft–led assisted migration, before this approach could even be tested with an endangered species like the Whooping Crane, it had to be tried and refined on more expendable forms of life. Sandhills—

along with a range of other birds, including Canada Geese and Trumpeter Swans (*Cygnus buccinators*)—were enrolled to do this difficult and dangerous work. Through a series of experiments conducted over roughly a decade, it was determined that major hazards included predation by Golden Eagles (*Aquila chrysaetos*) and collision with power lines and the propellers of ultralight aircraft. Many birds were injured or killed in these tests. In some cases, experimenters deliberately selected "hazardous" migration routes with hundreds of power-line crossings to thoroughly explore the nature of the potential danger and "trouble shoot solutions before [they] began with endangered cranes" (Ellis et al. 2001:141, 2003:262).

Other birds have also been drawn into this conservation project in less visible ways. At Patuxent, for example, a quail colony has been established to play the role of "royal tasters," testing all new batches of Whooping Crane feed to ensure their desirability and safety (this colony was established after an incident in which a new batch of food was contaminated by microtoxins that made Whoopers sick [French, pers. comm.]). Beyond the edge of these breeding facilities, the use of "sentinel turkeys" has also been proposed: captive or free-roaming birds placed at potential Whooping Crane release sites to determine the likely impacts of various diseases and so the suitability of the sites for the cranes (USFWS 2011:6074).

All these birds take on a kind of "sacrificial surrogacy" in which, by virtue of an actual or asserted similarity, one being is able/required to stand in for another and, importantly in doing so, to bear the cost or harm of that other (Reinert 2007). In this case, the relevant similarities are biological, behavioral, and/or social. It is because of these kinds of similarities that information, experiences, and the real physical labor of individuals of one species in domains like eating habits, toxic exposure, incubation, and migration are able to substitute for those of another: able to do the work of another or offer reliable and meaningful insight into another's life and possibilities. This is not an unusual situation. French (pers. comm.) sums it up simply: "Like with many endangered species programs, a strategy that we've taken is to do the experimentation on a closely related, hopefully congeneric, species. So you make your mistakes on the common species and then assume that the results that you get from those experiments can apply to the closely related endangered species." All these birds have been drawn into the plight of the Whooping Crane—into the captive life of a species at the dull edge of extinction—and, in one way or another, have posed to stress and potential disease. These two sets of practices are thoroughly interwoven. *Together,* they produce hope for the possibility of the continuity of Whooping Crane lives and lifeways.

When presented with this situation, perhaps many of us would still choose the violence of a conservation grounded in captive breeding over that of extinction. But making this decision cannot be allowed to erase this genuine ethical difficulty: the violence of the care that is practiced here. Instead, making a stand for conservation must require that we actively take on this predicament, that we consciously dwell within it, in an effort to, wherever possible, work toward something better. In Donna Haraway's (2011, forthcoming) terms, conservation in this complex space requires a dedicated and relentless practice of "staying with the trouble." Guided by this practice, my approach to the ethics of Whooping Crane captive breeding is not grounded in a universal system or generalizable principles, but is an attempt to inhabit the complexity and difficulty of real relationships in entangled multispecies worlds without the comfort of simple answers or justifications.

At the heart of the practice of "staying with the trouble" in Whooping Crane conservation—as well as in most other endangered species conservation projects—is the need to move beyond what Haraway (2008), following Jacques Derrida, has called a "sacrificial logic." This is the unique brand of "species-thinking" (Chrulew 2011a) that positions individual organisms of both the endangered species in question and numerous other species implicated in the drama as "killable" in the name of the greater good of conservation. Refusing this sacrificial logic is about insisting that these individuals *do* make ethical claims on us. The point is not that we can never kill or cause suffering in the name of conservation, but that the decision to do so—no matter how rigorously we justify it and weigh our options—should never leave us comfortable and satisfied. In this context, killing or causing suffering for conservation can often still be "necessary, indeed good, but it can never 'legitimate' a relation to the suffering in purely regulatory or disengaged and unaffected ways" (Haraway 2008:72). Even if the death of the last Whooping Crane is delayed for several more decades, even if the species recuperates to the point that it can again be self-sustaining in multiple free-living migratory populations, the violence and suffering that has produced this outcome will not and cannot be erased or "justified away."

As we have seen throughout this chapter, competing regimes of care overlap and sometimes come into conflict in the captive breeding of Whooping Cranes. The practical labor of caring for this endangered species has produced a set of approaches and protocols that sometimes undermine efforts to care for individual lives of various kinds. And so some forms of flourishing are sacrificed for the sake of others. This is an all-too-common occurrence in conservation projects where care for the species is usually uncritically taken to trump all other ethical concerns.[16] In this context, refusing a sacrificial logic requires us to develop ways of caring for cranes that take seriously the flourishing of both species and individuals. Doing so prevents us from inhabiting a space of moral comfort that sidesteps any of these genuine ethical demands. Here, we are exposed to an *imperative* to constantly strive to do better than we are now. *If* captive breeding is to continue—if, even under the weight of all this suffering, we feel that it is ultimately a "good"—then how might it be done in the best possible of ways?

The first thing that is required is an ongoing effort to improve the lives of cranes and all of the other species that have been drawn into this difficult space. The staff that I met at Patuxent, ICF, and Operation Migration cared deeply for their charges. Working within the constraints provided by budgets, personnel, and, most important, crane biology and development, they did a fantastic job, in many different ways, of rearing and providing for these birds. For the most part, however, the protocols for day-to-day life that dominate in captive breeding facilities like these are driven by functional outcomes. While practices like costume rearing and habituation to AI might improve the lives of captive and released birds, it was clear to me in interviews that both are motivated primarily by other priorities: increased reproductive success and more reliable (and less labor-intensive) breeding behaviors. If either of these practices did not have direct benefits for the conservation of the species, it seems likely that they would not have been adopted.

At Patuxent, John French discussed with me the possibility of experimenting with some parental rearing of Whooping Crane young. This would mean that instead of eggs being incubated by Sandhill Cranes or machines, and then chicks being raised by costumed humans, Whooping Crane parents would fulfill these roles for at least some of their young.[17] At the time of our discussion, this proposal had not been explored in great detail, but again the motivation was functional. While birds released into the Eastern Migratory Population were surviving, they did not seem to be breeding well. In fact, from the hundreds of birds released over roughly the past ten years, only three young have been successfully fledged. There are many theories as to why this might be; one of the dominant ideas is that a large number of black flies in the Necedah National Wildlife Refuge, in Wisconsin, are driving birds to abandon their eggs and young. It is for this reason that new release sites for the population have recently been adopted in Wisconsin. While black flies may well be the primary cause of a lack of breeding success, French is concerned that there is at least a chance that captive-reared birds may not be well equipped when it comes to rearing their own young. In part, this is because their upbringing has been so unconventional: instead of spending all their time in the company of their parents and rarely seeing another young bird, captive-reared chicks spend a lot of time in the company of other young birds, and the role of their parents is filled primarily by visits from costumed humans. In this context, there is at least a chance that parent-reared birds might be better prepared to raise their own young once released.

This is precisely the kind of change that might be made to improve the lives of captive cranes. As zoo- and lab-based research over the past few decades has shown, social and environmental enrichment of this kind plays a key role in improving the lives of captive animals (Hosey, Melfi, and Pankhurst 2009). In many European zoos, rearing young is thought to offer an important form of social enrichment. As Bengt Holst, director of conservation at the Copenhagen Zoo, put it: "We have already taken away their predatory and antipredatory behaviors. If we take away their parenting behavior, they have not much left" (quoted in Kaufman 2012). To this end, instead of utilizing contraceptives, as do many American zoos, the Copenhagen Zoo allows animals to raise their young up to the age when they would be expected to separate from their parents. At this point, given the crowded nature of many zoos, the young animals are "euthanized." I would like to set aside the many ethical issues that this practice raises, a practice that Marc Bekoff (2012) has argued is fundamentally different from euthanasia and should instead be called "zoothenasia." Instead, what is of central importance in this context is the way in which this practice highlights the potential enrichment benefits of allowing animals to be involved in rearing their own young. Studies on the diversity provided by enclosures, feeding methods, toys and puzzles, and other stimuli have provided a range of other options for enriching the lives of captive animals (Hosey, Melfi, and Pankhurst 2009:259–88).

Some enrichment practices have been taken up in the Whooping Crane program, and others are being explored; for example, a trial is under way at Patuxent to introduce some wetland features into some crane enclosures (French, pers. comm.). As at many intensive captive breeding facilities, however, where these kinds of enriching changes have occurred, they have tended to arise through the happy alignment of regimes of care for species and individuals. In contrast, a sustained and situated ethical engagement might ask us to *also* value, support, and invest in research and changes for the "simple" benefit of enriching individual captive lives. More and more might be done in those areas where caring practices align. Where they do not align, tough decisions will have to be made. But they must be made as *decisions*, rather than simply taking place under the assumption that the "good of the species" always trumps everything and everyone else.

In addition to opening up the scope of ethical consideration and working to improve captive lives, there is a third important characteristic to the ethical practice that I am advocating: making public the realities of this brand of emergency conservation in terms of the lives, suffering, and quality of life of a range of living beings across generations. At its heart, this is a practice of rendering visible some of the "abject critters . . . proliferating in the shadows of human dreams and schemes" (Kirksey, forthcoming). As it stands, the public presentation of captive breeding and release programs usually focuses on simple, predictably sensational, success stories. In this context, the Whooping Crane story is perfect, with its spectacle of ultralight-led migration and its uncomplicated redemptive framing: humans put technology to use to save birds and lead them on migration. This framing backgrounds the realities of captive life, covering them in a veneer of uncomplicated "care." In fact, it is often an idealized "motherly care" that is invoked in this context: Konrad Lorenz as "goose mother" and cranes following ultralight aircraft as they would their own mothers. It seems that when leading cranes on migration, even Russian president Vladimir Putin can be presented in this frame. Much of the cloud of media interest that accompanied Putin's "flight with cranes" for a similar program with the endangered Siberian Crane (*Grus leucogeranus*), in September 2012, utilized this predictable imagery: variously referring to Putin as "mother hen" and "mother crane" (Malein 2012; Stewart 2012).

In some conservation circles, too, captive breeding and release has come to be regarded as an increasingly popular option. While Whooping Cranes were certainly one of the early pioneers of this approach, over the past few decades it has been applied to the conservation of numerous other endangered species. The Indian vulture and the Hawaiian Crow are just two examples (chaps. 2 and 5). In all, Noel Snyder and his colleagues (1996) report that in the early 1990s, captive breeding was recommended by the International Union for Conservation of Nature for 50 percent of the world's parrot species, as well as in 64 percent of all approved recovery plans for threatened and endangered species in the United States. It seems that despite their huge financial costs and significant practical difficulties—including very low success rates (Bowkett 2009; Fischer and Lindenmayer 2000; Snyder et al. 1996)—captive breeding programs are still often viewed as a good option for endangered species (Martin 2012).

Juvenile Whooping Cranes in a holding pen awaiting release. (U.S. Fish and Wildlife Service/Southeast; CC BY 2.0)

But spending time with some of the creatures whom we have brought into these programs complicates this approach, along with the simple narratives of care and success that often surround it. The Whooping Crane story makes clear that extinction is far more complex than a yes-or-no, black-or-white phenomenon. More than life and death is at stake here, for both individuals and species: there are *modes* of living and not-quite-fully living, modes of death and dying. These are the spaces of a "wounded life" (Chrulew 2011a). By drawing the reality of captive breeding practices into the foreground, we highlight some of the other "costs" of this brand of emergency conservation, enabling us to begin to think more critically about what is at stake in bringing species into, and then holding them captive within, the dull space at the edge of extinction.

As I finalize this chapter, I have been periodically checking in on the six young Whooping Cranes that I met at White River Marsh a few months ago. As I live on the other side of the world, this checking in has been facilitated by a blog and a live-streaming "crane cam."[18] All six birds are doing well, although one has had a cough and seems to tire easily. It is hoped that before too long, all the birds will set off on their first long journey south. As in previous years, this journey will be sponsored by a combination of merchandise sales and donations. There is also little doubt that it will again capture the interest and imagination of local communities and be keenly observed and discussed as the birds pass overhead. Like all the other people involved in this project, I am hopeful for the future of Whooping Cranes. But I want also to be attentive to the particular ways in which this hope is produced. I want to encourage care for a species that has been brought to the edge of extinction; but I want also to acknowledge the, to some extent unavoidably, violent reality of the way in which much of this care for species and environments is practiced. This chapter is an attempt to advocate for hope and care of these kinds. Ultimately, I remain convinced that certain kinds of violence and suffering "should" continue, but that our modes of thinking need to be radically different—such that other, more ethical, ways of living together in captive spaces might become both *visible* and *imperative*. Dwelling within this complexity while refusing to accept ultimate justifications is the messy work of ethical conservation for our time.

Five

MOURNING CROWS

Grief in a Shared World

> I remember most of all the Ho'okena bird, how after it lost its mate it cried out for weeks . . . a terribly high-pitched sound, like an inconsolable moaning. . . . The Ho'okena bird is so obviously looking for company, but there is none to be found—nowhere.
>
> GLENN KLINGER, QUOTED IN MARK JEROME WALTERS, *SEEKING THE SACRED RAVEN*

Death, mourning, and that collective mode of dying called "extinction" are painfully drawn together in this short quote. The bird in question, now long dead itself, was a member of that rarest of corvid species, the Hawaiian Crow (*Corvus hawaiiensis*). At the time that biologist Glenn Klinger spoke these words, only three of these birds were left in the wild. A couple of years later, in 2002, the last sighting of a free-living Hawaiian Crow was made. Since then, the only surviving crows have lived in captivity, subjects of a long-running breeding and conservation program (USFWS 2009).

This chapter explores the plight of the Hawaiian Crow, but it does so through a very particular lens: mourning. Drawing on a broad range of material concerned with crow behavior and ecology, my interest is in learning more about how these birds mourn for the deaths of others of their kind. Alongside this discussion, this chapter also draws on a philosophical literature in an effort to explore what it might mean for us to mourn for crows in a time of extinctions. Taken together, these two acts of mourning point to the possibility of our learning to mourn *with* crows for some of the many losses of life and diversity that take place within our shared world.

A captive Hawaiian Crow at the Keauhou Bird Conservation Center on the island of Hawai'i. (Photograph by author)

But this chapter is not just *about* mourning. In addition, it aims itself to *be* an act of mourning: to tell stories about the dead and dying that draw them into relationship with the living. In doing so, this chapter attempts to work across and break down the human exceptionalism that, as we will see, has so often dominated our thinking about death and our relationships with other animals and the broader environment. It is in part this exceptionalism that holds us distant, intellectually and emotionally, from our more-than-human world. Mourning offers us a way into an alternative space, one of acknowledgment of and respect for the dead. In this context, mourning undoes any pretense toward exceptionalism, instead drawing us into an awareness of the multispecies continuities and connectivities that make life possible for everyone.

THE CROW THAT IS NOT A CROW

If you had traveled into the dense volcanic forests of Hawaii's Big Island a century ago, you may well have been lucky enough to catch sight of a Hawaiian Crow. In fact, you may not even have had to look very hard. Deeply inquisitive by nature, Hawaiian Crows seem to have frequently greeted early naturalists who made their way into the island's forests (Walters 2006). According to one of these naturalists, Henry W. Henshaw, in *Birds of the Hawaiian Islands* (1902):

> The bird, instead of being wary and shy, seems to have not the slightest fear of man, and when it espies an intruder in the woods is more likely than not to fly to meet him and greet his presence with a few loud caws. He will even follow the stranger's steps through the woods, taking short flights from tree to tree, the better to observe him and gain an idea of his character and purpose. (quoted in Walters 2006:63)

As is perhaps implied by the vivid image that Henshaw's words paint, the forest was central to the life of these crows. Although sometimes venturing beyond its borders, Hawaiian Crows lived primarily among the trees, relying on them for the invertebrates and forest fruits that made up the bulk of their diet (Banko, Ball, and Banko 2002). They even made use of the forest flowers, eating some whole while probing and piercing others in search of nectar. As the island's largest forest bird, and a largely frugivorous one at that, the species is thought to have probably played an important role as a seed disperser, "potentially influencing the composition and function of dry- and wet-forest ecosystems" (Banko, Ball, and Banko 2002).

Perhaps, in most of these habitat and dietary preferences, these birds do not really sound like "crows" at all. The broad crow family (Corvidae, often referred to as "corvids") is composed of many kinds of birds, including jays, magpies, ravens, and crows. But it is these last two types of predominantly black birds—crows and ravens, along with the Jackdaw (*Corvus monedula*) and Rook (*C. frugilegus*), sometimes collectively called the "true crows" (genus *Corvus*)—that most people think of when they hear the word "crow." While there are many species of "true crow" around the world, the ones that many of us know best—those that make their homes among us, living in cities and rural areas—are in many ways very different kinds of birds from those found in Hawai'i; we might think here of species like the American Crow (*Corvus brachyrhynchos*), the Australian Raven (*C. coronoides*), and the House Crow (*C. splendens*) in India and other parts of South Asia, not to mention perhaps the most successful member of the genus *Corvus*, the Common Raven (*C. corax*), a species that can now be found over fully half of Earth's landed surface (Marzluff 2005:47). All these well-known species are omnivorous and opportunistic, generalists of the most blatant kind. They are willing and able to live in a wide range of habitats and situations, exploiting a similarly wide range of food sources. Much of this diet—at least the bit most visible to people—is now often composed of scavenged waste, whether carcasses collected along roadsides or rubbish pulled from bins or dumps. It was with these kinds of crow species in mind that biologist John Marzluff (2005) noted, "If crows can be thought of as specialists in any way, they are specialists on people" (32).

But this urban scavenging lifestyle has often earned crows little fondness in people's hearts. This situation was, in fact, part of the motivation behind conservationists' decision to refer to the crow by its Hawaiian name, 'Alalā, thus emphasizing its considerable differences from many more well-known corvids and undoubtedly helping in efforts to raise funds for and public concern about the future of the species (Lieberman, pers. comm.).[1] As a fruit and forest specialist, the Hawaiian Crow is already very different from a lot of other crows. But, importantly, it is also unlike many of these other species in terms of its response to human habitation. Whereas many other corvids have thrived in company with humans, the Hawaiian Crow has instead been driven to the very edge of extinction.

The key problem for the Hawaiian Crow, as with so many other island birds, has been rapid and ongoing alteration of the environment. Hawaiian birds have had to survive through two waves of significantly different human settlement occupation: first the arrival of Polynesians, about 1,500 to 2,000 years ago, and then the arrival of Europeans beginning in the late eighteenth century. In each case, many species have been lost. Today, Hawai'i has the dubious honor of being home to more endangered species per square mile than any other place on Earth (Restani and Marzluff 2002; Steadman 1995). While Hawai'i is undoubtedly a particularly bad case, small islands all over the Pacific—and, indeed, around the world—are in a similar position. As Marzluff (2005) notes, "In little over a thousand years we have extinguished more than half of all the bird species that occupied the lush islands of the tropical Pacific" (256).

The environmental change that has all but wiped out the Hawaiian Crow has taken a variety of forms. At the most obvious level, the loss of large areas of forest has decreased the possible range of the species, while also reducing the availability of some food plants. These transformations have been incredibly widespread. As the U.S. Fish and Wildlife Service's (2009) recovery plan for the species notes: "There is no existing forest within the historical range of the 'Alalā that has not been substantially altered from its pre-European condition, much less from its condition prior to the [human] colonization of the islands" (I-10). In addition, the introduction of a range of animals to Hawai'i has produced new predators for crows, while increasing vulnerability to existing predators. Newly arrived species like rats, mongoose, and cats attack crows and their eggs, while pigs, cattle, and other grazing animals have thinned out the understory in surviving forested areas, making crows more vulnerable to predation by the 'Io (Hawaiian Hawk [*Buteo solitarius*])—a species that is itself listed as endangered. In addition, humans have played a role as direct predators of crows, with farmers in the past even taking advantage of these birds' curiosity by imitating their calls to attract and shoot them (Marzluff 2005:259; Walters 2006:62). Alongside all these threats, introduced diseases—in particular, toxoplasmosis, avian malaria, and avian pox—have likely taken a huge toll on crows and a range of other birds, either killing them outright or significantly weakening them for other predators.[2]

Today, the Hawaiian Crow is extinct in the wild. Reduced to a mere 100 birds in captivity, it is widely thought to be the most critically endangered corvid on the planet (Banko, Ball, and Banko 2002:25).

Despite the significant differences between the Hawaiian Crow and many of its corvid cousins—differences in terms of habitat and diet—the Hawaiian Crow is most definitely crow-like in other important ways. In particular, Hawaiian Crows likely share with other corvids a high degree of intelligence and a capacity for deeply social and emotional lives. Corvids are clearly among the most intelligent birds, and perhaps animals more generally. So intelligent, in fact, that Nathan Emery (2004) has proposed—in the hominid-centric language typical of much human thought on intelligence—that corvids might reasonably be thought of as "feathered apes," a view shared by many other biologists (Emery and Clayton 2004; Heinrich and Bugnyar 2007; Marzluff 2005:40; Seed, Emery, and Clayton 2009). This is the intelligence that Henry Ward Beecher surely had in mind when he made the now oft-quoted statement: "If men had wings and bore black feathers, few of them would be clever enough to be crows." Whether it be Carrion Crows (*Corvus corone*) in Japan who have learned to use traffic lights and moving cars to open tough nuts (Marzluff 2005:240) or New Caledonian Crows (*C. moneduloides*), another forest-dwelling island species, with their unique local cultures of tool construction and layers of meta-tool use (Hunt 1996; Taylor et al. 2007), corvids have, again and again, shown themselves to be highly intelligent.

Over the years, a range of experiments concerned with the intellectual, social, and emotional complexity of corvid lives have pointed to highly sophisticated abilities previously thought absent in the avian world, and perhaps restricted to humans, other primates, dolphins, and a select few other animal species. In terms of cooperation and the coordination of behavior, the consolation of mates after conflict, self-recognition, and the attribution of mental states to others, corvids seem remarkably good at understanding and interacting with one another and the wider world (Bugnyar 2011; Bugnyar and Heinrich 2006; Fraser and Bugnyar 2010; Pika and Bugnyar 2011). This does not mean that relationships are always perfectly amicable inside a murder of crows, or even for a mated pair, but it undoubtedly means that crows lead cognitively and emotionally rich lives (at least in the terms that matter enough to humans for us to measure them).[3]

Throughout this chapter, I have drawn on a large ethological literature on a range of corvid species, primarily studies concerned with true crows. In doing so, I hope to reinforce the link between the Hawaiian Crow and the broader corvid family. While I understand conservationists' decision to refer to the species as 'Alalā—and perhaps to downplay this familial connection—I will instead use the name Hawaiian Crow throughout. My hope is that by the end of this chapter, having been drawn a little further into the world of this remarkable avian family, readers will consider the word "crow" in the positive light that it deserves (if they do not already). In addition, however, I have drawn on this broad corvid literature for the simple reason that we just do not know very much about the Hawaiian Crow specifically, and the species is now far too endangered to become the subject of relevant experiments. There is, however, every reason to believe that these crows share the cognitive and emotional attributes found among the other members of their genus—attributes that, it should be noted, have been shown to be possessed by both those generalist species with whom most of us live and a number of other island corvids with more specific dietary and habitat requirements (for example, the New Caledonian Crow). And so in drawing on this general corvid literature, my aim is to also say something as concrete and realistic as is now possible about the experiential world inhabited by Hawai'i's crows, while exploring what we might learn about death and grief in a time of extinctions from thinking and mourning with corvids.

DEATH AND HUMAN EXCEPTIONALISM

Before turning to the crows themselves, however, there are some more general points on philosophy, death, and the nonhuman that deserve mention. Although taking a variety of different forms throughout the history of Western thought, the role of animals in thinking about death—as in so many other contexts—has almost always been as a foil for thinking "the human." Indeed, it seems fair to say that whenever animals have been mentioned in the same breath as death in Western philosophy (at least until very recently [for example, Plumwood 2002]), it has inevitably been to distinguish something unique about human knowledge or experience, something that sets us apart from the rest of the animal kingdom.

A central strand of thought has been the long-standing and widespread assumption that nonhuman animals do not "know death." Voltaire tells us that "the human race is the only one that knows it must die," while according to Schopenhauer, "Unlike man, animals, so to speak, live without knowing death" (quoted in Enright 1983:iv). In the twentieth century, this idea found perhaps its most ardent and eloquent support in the work of Martin Heidegger, according to whom the animal cannot "die." While, like all living things, animals will inevitably come to an end or "perish," for Heidegger (1996:246–49) humans are unique in our relationship with that ending, in our ability to be consciously oriented toward our deaths—in his terms, to "die."[4] This distinction is utterly central to Heidegger's larger philosophical project (Calarco 2002; Derrida 1993). The notion that the animal cannot die interacts in his work in a complex manner with a range of other ideas about how animals differ from humans—the animal has no language, has no "hand," is "poor in world"—each of these ideas informing and reinforcing one another in a way that ultimately yields a picture of humans as thoroughly and *essentially* different from the rest of the animal kingdom (Buchanan 2008:45). Ultimately, as Matthew Calarco (2008) has argued, Heidegger's work in this area is an attempt to chart an absolute and abyssal divide between "the Being of human beings and that of [other] animals . . . a gap and a rupture that is utterly untraversable" (22).[5]

French philosopher Françoise Dastur (1996) has in a way parted company with this long tradition of thought, moving away from a singular focus on the individual's own death—what she calls a "phenomenology of mortality" (42)—toward a more relational account centered on our memories of, and interactions with, the dead. This is a promising movement, and one that influences my thinking here. And yet, there is something profoundly traditional and humanist about the way in which Dastur takes up this idea; despite her shift in focus, death still emerges from her work as the basis of a divide between the human and the animal: "That human life is a life 'with' the dead is perhaps what truly distinguishes human existence from purely animal life" (Dastur 1996:8). Dastur's thought seems to center on the notion that the political and cultural dimensions of human life inevitably "reference" the dead: whether directly, in the sense that the dead continue to live among us and act on us as spirits or ghosts, or "simply" in terms of the meanings, values, memories, and ideas that we individually and collectively inherit (not to mention the languages and other modes of expression that we inherit them through). In this context, all human life takes place among the living *and the dead*: a person "lives in society not merely with his 'contemporaries' but also—and perhaps more so—with those who have gone before" (Dastur 1996:8).

Death does important boundary work in this kind of philosophical thought. Knowledge of death, or a relationship with the dead, here joins a long list of other "lacks," other characteristics or attributes that are thought to ground an essential difference between humanity and animality: be it the possession of language, mirror self-recognition, rationality, moral agency, or any number of other characteristics (Calarco 2008:75; Haraway 1989). In this context, death has become another of the "propers" that Dominique Lestel (2011) has described: an essential and unique characteristic *proper* to the human that does not just make us different in the way that all animal species are different from one another, but somehow sets us outside the sphere of animality. In Val Plumwood's (2007) terms, these ideas about death form another important site for the development of philosophical modes of "human exceptionalism"—that is, "the idea that humankind is radically different and apart from the rest of nature and from other animals."

But, as with all these other supposed "lacks," it is far from clear that death is up to the task of dividing up the animal kingdom so neatly or finally. If we take seriously *specific* nonhumans and the current scientific literature about them, examples abound of animals interacting with the dead in ways that, at the very least, must draw us to question these ideas: from foxes burying others of their kind and gorillas caught up in obvious displays of profound grief (Bekoff 2007:63–65), to the long periods of interaction with the bones of the dead that so often occur in elephant communities, sometimes covering them with leaves or branches, and at other times slowly and silently touching them with their trunks and feet (Poole 1996:153–55). Reading longer accounts of these behaviors, it is often hard to believe that these animals do not have some notion of death, some concept that the other is no longer with them in the same way and will not be again. What else might it mean to a fox or an elephant to bury or cover the body of another or to return to their bones, again and again?[6]

At the same time, in their frequent return to touch the bones of their dead, I see in elephants' grief a quiet but profound challenge to Dastur's

(1996) notion of the human. Can we possibly think that elephants do not dwell with the dead in their own elephant ways? Can we really believe that their community, their lives, are not also structured around and lived in reference to those who are now gone? The captive Hawaiian Crows who are thought to have lost parts of their vocal repertoire in the absence of free-living adults to learn from (Lieberman, pers. comm.) also offer a powerful example of a nonhuman community that has traditionally drawn from and referenced those who are no longer living corporeally among them. In fact, in a whole host of ways, many nonhuman animals "reference" those that are now dead. In *learned* behaviors from hunting and defending territory to singing songs, these animals inherit and adapt ways of life that are profoundly shaped by previous generations.[7] As is suggested by the plight of Hawai'i's crows—alongside that of elephants around the world, many of whom are endangered and subject to ongoing anthropogenic violence—perhaps the ability to live in a way that references and interacts with the dead is not uniquely human *as such,* but rather is a way of life that we are increasingly denying to a host of other animals.[8]

Ultimately, my intention in this chapter is not to directly refute these ideas about who "knows death." In the final analysis, this is an issue that cannot be resolved with certainty. In fact, as Jacques Derrida (1993) has noted, it is far from clear that we humans "know death" or, for that matter, even know what it would mean to "know death." Instead, my goal is simply to tell different stories about death: to shift the focus from knowledge of death to the experience of grief at its occurrence and the possibilities that this grief may open up for crows, humans, and others. Instead of reproducing a human exceptionalism that separates us from the rest of the world in yet another way—as traditional philosophies of death have tended to do—my goal in taking up this focus is to explore some of the many ways in which death entangles us in multispecies worlds.[9]

KEEPING COMPANY WITH CROWS: THE EVOLUTION OF GRIEF

This chapter opened with a Hawaiian Crow crying out—seemingly in grief—at the death of its partner: the "high-pitched . . . inconsolable moaning" that one of the scientists working on their conservation described

(Klinger, quoted in Walters 2006:241). There are, however, relatively few accounts in the scientific literature of crows grieving, and those that do exist tend to be anecdotal.[10] Perhaps most famously, in his popular book on animal behavior, *King Solomon's Ring*, Konrad Lorenz ([1949] 2002) tells the story of a mourning Jackdaw. The bird was the last remaining member of a group of Jackdaws kept by Lorenz that mysteriously disappeared from his aviary; the other birds likely either escaped or were killed by a predator. Lorenz recounts that

> [her song] was really heartrending. It was not how she sang, but what she sang. Her whole song was suffused with the emotion which obsessed her, with the sole desire of bringing back her lost ones by means of the "Kiaw" call, "Kiaw" and again "Kiaw" in all tones and cadences, from the gentlest piano to the most desperate fortissimo. Other sounds were scarcely audible in this song of woe. "Come back, oh, come back!" (163–64)

In an interview, biologist John Marzluff (pers. comm.) also recounted witnessing behavior among American Crows that looked like grief: "I've seen birds grieving, I think it would be fair to say. I've seen a bird that was perched above another bird that was dying—I don't know if they were related or mated, or what—but, certainly that bird was very attentive and watching that other bird on the ground as it was dying."

Over the past few decades, these kinds of observations—now made in relation to a range of animals, including elephants, some primates, and crows—have driven a growing acceptance that some nonhuman animals experience grief at the deaths of others (Archer 1999; Bekoff 2007).[11] While theories differ slightly, at the core of the story that many evolutionary psychologists and ethologists now tell about grief is the notion that it is intimately entangled with the evolution of close social relationships, which are themselves desirable for the many fitness advantages that partner/group living can confer on individual existence (Archer 1999). I am partial to the influential version of this account first offered by Colin Murray Parkes, who saw feelings of love for others as essential to the maintenance of close relationships, and consequently understood grief as the "cost of commitment": the cost of this evolved ability to relate and *be* meaningfully with others (Archer 1999:60).[12] While there are obviously differences in the forms that this grief takes within and between species

One of the last free-living Hawaiian Crows, at McCandless Ranch on the island of Hawai'i, March 1998. (© Jack Jeffrey Photography)

of animals, as John Archer (1999) has noted, there is now "abundant evidence that reactions essentially similar to those shown by humans occur in social animals which [who] have lost or been separated from a social companion" (55).

Crows are highly intelligent and social birds, and are perhaps stereotypical of the species that we might expect to have evolved a capacity to grieve. Support for the existence of grief in crows is perhaps also offered by their expression of related emotional states that are connected to their close social relationships. As with grief, it has been argued that empathy likely evolved in animals as a result of selection pressures in social environments. As Frans de Waal (2008) has succinctly put it: "Empathy allows one to quickly and automatically relate to the emotional states of others, which is essential for the regulation of social interactions, coordinated activity, and cooperation toward shared goals" (282). According to him, the relevant selection pressure probably originated in the context of parental care, in which those birds and mammals that were "alert to and affected by their offspring's needs likely out-reproduced those who remained indifferent" (de Waal 2008:282). "Empathy" takes many forms here, ranging from a simple awareness or sharing of another's emotional state (emotional contagion), through to more complex forms of understanding and targeted helping.

Recent studies on both Rooks and Common Ravens have revealed behaviors that point strongly to a highly developed capacity for empathy (Fraser and Bugnyar 2010; Seed, Clayton, and Emery 2007). These experiments found clear patterns of "postconflict affiliation" among their subjects. This affiliation was limited to mated pairs and took the form of close physical interaction after one member of the couple had been involved in a conflict with another bird. There are various possible explanations for this behavior in different species. On the basis of their findings with Common Ravens, however, Orlaith Fraser and Thomas Bugnyar (2010) have argued that this behavior likely plays an important conciliatory role, comforting and supporting a distressed partner. Consolation requires a "cognitively demanding degree of empathy" in which a bystander must "first recognize that the victim is distressed and then act appropriately to alleviate that distress" (Fraser and Bugnyar 2010:1; see also de Waal 2008:285–86).

The link between grief and empathy is not at all straightforward. But in the absence of more documentation of grief, these kinds of empathic responses seem to point to the kind of emotional and social entanglements—the *kinds* of shared lives, the *being at stake in each other's company*—that would likely give rise to grief at the severing of a social bond, especially a pair-bond in the case of crows.

In thinking in these ways about emotions like grief and empathy, we do productive work in undermining human exceptionalism by drawing our own responses to death into an evolutionary continuum. While Darwin's work should rightly have put human exceptionalism of this kind to rest long ago, as Val Plumwood (2007) has noted there is an important sense in which evolution has often been interpreted in a way that simply shifts the site of our exceptional nature. While *bodily* continuity with the animals is now readily acknowledged, "[t]he radical break or discontinuity that characterizes exceptionalist thinking has not been abandoned with modernity, but has been located elsewhere—in the human mind" (Plumwood 2007). While some emotions have been conceptually placed on the bodily side of this dualistic division—and there is a long history in the West of thinking about emotions as base bodily impulses (Despret 2004b:37–38)—other more "complex" emotional states, such as grief, have usually been linked to "developed" cognitive capacities and, as a result, have been taken to be the exclusive possessions of *Homo sapiens* (Bekoff 2006). In this context, paying attention to the *evolution* of grief goes some way toward unsettling this fallback exceptionalist position. While emotions like grief certainly take myriad forms among the many social mammals and birds, they are nonetheless shared in an important sense, too (as is increasingly being shown in work on the neuroevolution of empathy and other emotions [Decety 2011]). Darwin, of course, knew something like this when he located the roots of human grief, as with the rest of our emotional repertoire, in the animal world in *The Expression of the Emotions in Man and Animal* ([1872] 1965; Crist 1999:17–29).

In addition, paying attention to mourning crows enables us to understand a little better the experiential world that Hawaii's crows inhabit—at least for now. In doing so, we gain a "thicker" sense of who these creatures might be, but also of what is being lost in their disappearance. Far more than "biodiversity" in any narrow sense, mourning crows remind us that whole modes of life, whole ways of living and dying in company with others, are disappearing—nonhuman languages, socialities, perhaps even cultures.[13] Part of this loss will inevitably also be *ways of mourning*. Perhaps in the end, what must be mourned at this time, alongside so many other things, is the diminishment of mourning itself, the loss of the rich and varied expressions of grief that have evolved on this planet over millions of years. As species disappear, or as their socialities become dislocated and fractured by violence and disturbance, their ways of being meaningfully together in death, as in life, are undermined and lost (Rose 2008).

MOURNING AS RELEARNING A SHARED WORLD

But I suspect that corvids have still more to teach us about death and mourning. In exploring this possibility, we might start with a funeral. While traveling through the mountains in Colorado, ethologist Marc Bekoff (2007) witnessed a gathering of magpies in which four of these birds were standing around a fifth, likely killed by a car: "One approached the corpse, gently pecked at it . . . and stepped back. Another magpie did the same thing. Next, one of the magpies flew off, brought back some grass, and laid it by the corpse. Another magpie did the same. Then, all four magpies stood vigil for a few seconds and one by one flew off" (1). It is far from certain what these interactions may have meant for those birds on that day, or how widely similar practices might exist among other species—although since the publication of this account, Bekoff (pers. comm.) has been sent numerous reports of similar behavior among other corvids. To my knowledge, no such funerals have been observed among Hawaiian Crows, and perhaps we will never know for certain in what ways they marked the deaths of so many of their kind in recent decades. Although perhaps the experience of the Hoʻokena bird referred to in the epigraph gives us an important indication.

John Marzluff (pers. comm.) has also frequently encountered large gatherings of crows and ravens at sites of death. On several occasions, he has even orchestrated these assemblies by placing a dead American Crow—one found that way—back in the environment:

> In all those cases—I've done it several times—their response was the same. The birds come in; they see the dead bird; they immediately fly down and start scolding. They will land around that bird and make a lot of noise and scold. And then, being gregarious animals, they'll probably start preening and doing lots of other things, and then eventually they fly off. Personally, I think that that's what everybody sees when they say they've seen a funeral. Basically, what's going on there is that the birds are learning about a very dangerous situation. . . . They're learning this is a dangerous place, or there is a dangerous predator, or some situation here that we need to know about and avoid in the future.

Death functions as a powerful stimulus to learning in this account. Marzluff's observations also indicate that the lessons to be learned from death are very quickly taken up by crows.[14] In fact, American Crows have been known to avoid places where one of their kind has been killed for over two years, sometimes changing whole flight paths to avoid flying over such a site (Marzluff, pers. comm.).

Clearly, crows learn about danger from death, but this fact in no way undermines the possibility that they may also experience grief at such times. In fact, if death does provide an important opportunity for learning, this outcome would only be enhanced by a strong emotional response, be it fear or grief. And so this possible evolutionary *function* for crow gatherings at sites of death does not, of course, mean that this is also the *motivation* of individual birds in attending.[15]

In pointing to this potential learning opportunity, however, Marzluff and the crows that he knows remind us that there is more to mourning than the "simple" expression of grief. In addition, as many psychologists and philosophers have insisted in relation to human grief, processes of individual and collective mourning do important work in allowing us to learn from and "work through" experiences of loss (Freud 1917; Riegel 2003). This idea has been expressed in a range of ways, but I am particularly drawn to philosopher and counselor Thomas Attig's (1996) understanding of grieving as a process of "relearning the world":

> As we grieve, we appropriate new understandings of the world and ourselves within it. We also become different in the light of the loss as we assume a new orientation to the world. As we relearn, we adjust emotional and other psychological responses and postures. We transform habits, motivations, and behaviors. . . . Some of what we took for granted in ourselves or in our life patterns is no longer viable or sustainable. Relearning the world thus requires that we make changes. (107–8)

In short, one of the core components of the way in which Attig understands grieving is as a more or less conscious process of learning and transformation to accommodate a changed reality.

What grief points to here is a particular kind of *shared* world or *shared* life. This is a way of being with others that, as far as we know, is unique to some mammals and birds, a particular sociality rooted in our being *emotionally* at stake in one another's lives. This possibility, this way of being with others, is a complex biosocial achievement, requiring the coming together of evolutionary histories and emotional and cognitive competencies to produce embodied subjects who are unavoidably emotionally entangled with one another.[16] It is only inside these particular biosocial configurations that the passing of another out of the world can be experienced and felt as a genuine loss. But loss is not experienced in the face of all change or even death. It is not enough for two such beings to have lived alongside each other, in proximity to each other; rather, they must also in some way have become *at stake in each other*, bound up with what *matters* to each other. In other words, they must in some sense, more or less consciously, have come to inhabit a meaningfully *shared world*.[17]

Grief, then, in Vinciane Despret's (2004a) terms, is a very particular process of "learn[ing] to be affected" (131), in which the borders between self, world, and other are profoundly problematized (209). This does not mean, however, that there is some sort of "default state" in which we are unaffected by the world, to which we must later add an emotional life. There is no default, originary position; there is only becoming-together inside rich histories of biosocial inheritance and relationship. In this context, learning *not* to be affected is equally a state that is produced: achieved through the cultivation of some relationships, some histories and understandings, and not others (Despret 2004a). As anthropologist Matei Candea (2010) makes clear in a somewhat different context, "ignoring" one another is neither a simple nor an originary mode of being with others for social animals attuned intellectually and emotionally to their complex surrounds. Rather, like engagement and attachment, it must be actively achieved (Candea 2010; Haraway 2008:24–25).[18]

It is with this understanding of grief in mind—as a complex biosocial achievement—that I would like to consider the general lack of popular interest in the deaths of species such as the Hawaiian Crow, which has become an all-too-common feature of our twenty-first-century world. What does it mean that, in this time of incredible loss, there is so little public (and perhaps also private) mourning for extinctions? Why do the last expressions of so many species leave this world unnoticed and unmourned—except perhaps by the few conservationists on whose watch and sometimes in whose hands, they pass away? (The others of their own kind being already gone, and so unable to mourn even if they once did.)

At the core of the answer that I would like to propose to these questions is our inability to really *get*—to comprehend at any meaningful level—the multiple connections and dependencies between ourselves and these disappearing others: a failure to appreciate all the ways in which we are at stake in one another, all the ways in which we share a world. This failure is, at least in part, rooted in the human exceptionalism that this chapter has explored. As Val Plumwood noted repeatedly throughout her long career, this kind of anthropocentric engagement with the world has important negative consequences for both humans and the many other living things that we share this planet with. As she put it in an important posthumously published paper:

> When we hyperseparate ourselves from nature and reduce it conceptually, we not only lose the ability to empathise and to see the non-human sphere in ethical terms, but also get a false sense of our own character and location that includes an illusory sense of agency and autonomy. So human-centred conceptual frameworks are a direct hazard to non-humans, but are also an indirect prudential hazard to Self, to humans, especially in a situation where we press limits. (Plumwood 2009:117)

The current anthropogenic extinction event is clearly one of those situations in which we are ever more dangerously pressing up against the limits of resilience of various ecosystems.

In Plumwood's (2007, 2009) account, human exceptionalism is positioned as doubly problematic. In the first instance, it is implicated in the erasure of the significance of nonhuman others, in our inability to empathize with their suffering and mourn for their deaths and ultimate extinction at our hands. As a dominant cultural narrative in many parts of the world, this is particularly so. The stories that we live by (Griffiths 2007)—as individuals and as societies—powerfully shape our ability to be affected by others (Despret 2004a:140). As these stories are taken up and lived, they "rearticulate" us as beings at stake in one another's lives in various ways. The affective separation of human exceptionalism holds the more-than-human world at arm's length: human exceptionalism plays a central role in the active process of our learning *not* to be affected by nonhuman others.

At the same time, however, Plumwood is attentive to the way in which human exceptionalism grounds a dangerous illusion in which the loss of nonhuman others is understood to never quite touch human lives and possibilities. No meaningfully shared world can emerge inside this conceptual space, and so the potential impacts of the loss of Earth's diversity on our own prospects for sustainable and meaningful lives are never quite grasped. As a result, we seem to have missed the real need for change—the need to relearn the world and our place in it—that death and grieving so often announce. As Marzluff's (pers. comm.) crows remind us, it can be very dangerous not to pay attention and make changes to behaviors in this context. But if the death of a single crow signals "here lies danger"—a danger significant enough to avoid a place for years, to alter flight ways and daily foraging routes—then what must the death of a whole species of crow, alongside a host of others at this time, communicate to any sentient and attentive observer? How could these extinctions not announce *our* need to find new flight ways, new modes of living in a fragile and changing world?

STORIED-MOURNING IN A TIME OF EXTINCTIONS

My hope is that this chapter about grieving crows may itself function as a narrative form of mourning. As Paul Ricoeur (2007) notes, "[T]he work of narrative constitutes an essential element of the work of mourning" (8). But this is so not just in the sense that stories help us to move on, to bear or even accept irreparable loss. In addition, stories play an important role in *communicating* this loss more widely, while helping to tease out the various ways in which loss matters, sometimes drawing distant listeners into a sense of felt connection and so affective involvement in a loss. A key part of this process is the "fleshing out" of the dead that stories enable, the chance to capture and communicate a fuller notion of who has died and why they mattered—in Judith Butler's (2009) terms, "to put together some remnants of a life, to publicly display and avow the loss" (39). In doing so, mourning may be an act of bearing witness to the deaths of so many individuals and species at this time in Earth's history. While some species may yet make it back from the edge of extinction, many others have not and will not in the years to come. In this context, mourning is a "simple" act of respect for and fidelity to those who have died.

But as they travel, stories also breathe new life into the dead, keeping them moving and enabling them to "haunt" our lives and future possibilities. In this sense, storied-mourning does not attempt to recover and move on from a loss—to put the dead to rest—but, as Jacques Derrida (1994) has suggested, offers us the possibility of mourning as a deliberate act of *sustained* remembrance that requires us to interrogate how it is that we might "live *with* ghosts" (xviii; Brault and Naas 2001; Ricciardi 2003). As Tammy Clewell (2009) has put it, what is at issue here is "a form of anti-conciliatory and sustained grieving that seeks to promote new [bio]social constellations so that the replaying of traumatic effects and injurious histories [and presents] might be shorn of their deadly consequences" (18–19). This is the kind of mourning that asks us—that perhaps demands of us, individually and collectively—to face up to the dead and to our role in the coming into being of a world of escalating suffering, loss, and extinction.

While there is potentially a kind of respect and acknowledgment in this refusal to put the dead to rest, there is also an important sense in which the dead are "put to work," a kind of "use" of the dead that Derrida (2001) has frequently cautioned against as an unethical (but, to some extent, also unavoidable) facet of mourning. And yet, as Derrida (1994) also acknowledges, "we know better than ever today that the dead must be able to work. And to cause to work, perhaps more than ever" (120). The work to be done here is, first and foremost, the task of "getting it" that these deaths, of individuals and of species, *matter*; that the world as we know it is changing; and that new approaches are necessary if life in its diversity is to go on. In this context, learning to mourn extinctions may also be essential to our and many other species' long-term survival.

It is not yet clear whether crows will make their way back into Hawaii's forests. The conservationists with whom I have spoken are hopeful, but also realistic about the many challenges that the species still faces—in particular, the need to restore forests and find ways to better protect birds from disease and predation. While I sincerely hope that this story has as happy an ending as is possible, this will not change the fact that countless birds have died and grieved, and that many generations of their kind will now be required to live in captivity for the species to have any hope of a future at all. Meanwhile, I cannot help but think of the literally hundreds of other species of Pacific birds who have already disappeared, just one part of our global impact on the diversity of forms of life over the past few human generations.[19]

It is in this context, inside deep histories of co-evolved affective bodies, that we are invited to mourn not just *for* crows, but *with* them. Hawaii's crows remind us that *if* we manage to find our way into a space of grief at this time, we will be just one species mourning among many, just one of the many forms of life on this planet that are experiencing this time of incredible loss through a lens of sadness and grief. In this context, mourning

EPILOGUE

A Call for Stories

In January 2013, while finishing work on this book, I returned to Hawai'i to continue my research on the Hawaiian Crow. While all these crows currently live their lives in captivity, it is hoped that in 2014 some of them may be able to be released. If this were to happen, and if those birds could form sustainable free-living populations, then a great achievement would have been made: forests that for over a decade have not heard the raucous calls of crows, or felt the movement of their graceful half-jump–half-flight through their canopies, would again be enlivened by this most charming and charismatic of birds. Beyond the pleasure that this return to the forest would bring to the crows themselves and to their observers (like me), it is likely that others would benefit, too. As the islands' largest frugivorous bird, the crow's absence may be having an impact on a number of tree species that have relied on these birds to disseminate their seeds (Culliney 2011). This is not an uncommon situation. When animal species disappear, the plants that co-evolved with them often feel this loss—some of them perhaps becoming extinct themselves in the absence of pollinators or disseminators (Barlow 2000; Janzen and Martin 1982). Speculation continues about a similar dependency of the tambalacoque tree on the now long-absent Dodo (Livezey 1993:272; Temple 1977).

But alongside endangered plants and their advocates (who might have much to gain from a return of the crow), on the Big Island of Hawai'i I also found many vocal critics of this proposed release. Foremost among them were pig hunters, concerned that the management of release sites for crows (in state-managed forests) would require that pigs be excluded and killed, limiting the number of animals that might be available for hunting. The relationships between humans and pigs in Hawai'i are complex: stretching from the arrival of the first people who brought small Polynesian pigs with them, through long histories of cohabitation and environmental change in which pigs have been central characters in the transformation of forests and the loss of numerous species of birds (Leonard 2008), to ongoing and sometimes heated contemporary tensions between conservation, on the one hand, and practices of pig hunting, on the other (Culliney 2011; Juvik and Juvik 1984).

All these entanglements are at stake in the conservation and possible extinction of the Hawaiian Crow. Will those tree species needy for crows for seed dispersal survive and thrive once again, or will they go the way of so many other 'anachronistic' plants? What kinds of human–pig cohabitation will be possible in the future to come? Will crows and many other endangered species find a place in the islands' forests, and at what cost to whom? Stretching back into the distant past and rippling forward into the many possible futures for Hawaii's forests, these are the kinds of complex relationships that characterize life at the edge of extinction. Much is at stake here for crows, but for a host of others, too.

This book has explored some of these kinds of entanglements in an effort to develop a broader, more complex, notion of what extinction *is*, and why and how it matters. This is necessarily a project that works against simplistic human exceptionalisms to tell stories that implicate us all—to varying extents and in a range of ways—in this incredible period of loss.

A motley gaggle of birds have been our guides. Each in its own way, they have required us to rethink what it means to be a fleshy, mortal creature, bound up with others, in a time of extinctions. In short, we have seen that what is at stake here are *ways of life*: ways of being with others, of mourning, of relating to a place, of rearing young, of making one's home in the world. All this stands to be lost as unique species slip out of the world, as millions of years of intergenerational labor, of evolutionary achievement, disappears. The natural sciences enable us to give some sort of an account of these ways of life: who these birds are and the experiential worlds that they inhabit, how they evolved, how they are woven into ecosystems with others.

But the natural sciences also need the humanities. This is the domain of the "environmental humanities," of a thinking that inhabits complex multispecies worlds without the aid (and impediment) of simplistic divisions between the human and the nonhuman, the cultural and the natural. The world is far messier and more interesting than this. And so the tools of ethnography and philosophy are required to develop a fuller picture of the entangled significance of extinction, of its myriad *meanings* and the diverse ways in which it *matters*. Alongside endangered species themselves, again and again we have seen that possibilities for ongoing life for a variety of others are drawn into extinction events: the loss of healthy environments to live in, of pollinators, of livelihoods for some and religious practices for others.

In taking seriously the entanglements of ways of life across evolutionary, ecological, affective, and multiple other domains, we are inevitably drawn into a set of complex *responsibilities* for what has come to pass and what may yet still be possible. If this period of incredible loss cannot rouse in us an awareness of our place in, and our responsibility for, a *shared* world, then I am not sure what can. The time has long since passed to learn a genuine appreciation for other forms of life, including the countless "animal subjects" (Noske 1989) with whom we share this planet, each with its own unique ways of inhabiting richly storied worlds.

This book is an effort to tell these kinds of stories. As noted in the introduction, extinction stories that implicate humans have a long history. But, despite this fact, we have not yet found good enough ways of thinking through what extinction *is* and what it *means*. At the same time, we have seen that there is no singular extinction phenomenon. Rather, in each case a different way of life, a different set of relationships and entangled significances, is at stake. And so just how these extinction stories might, or should, be told requires continual rethinking. Again and again, we need to ask: What does it means to bring an abrupt ending to *this* particular way of life? What does *this* loss mean inside its specific multispecies communities? How are "we" called into responsibility *here and now*, and how will we take up that call?

NOTES

INTRODUCTION

1. Just how large they were—or, rather, how "fat"—remains a topic of contention (Angst, Buffetaut, and Abourachid 2011a, 2011b).

2. According to Julian Hume (2006), the notion that Dodos were unpalatable is probably the result of a misunderstanding of reports that noted that they were less sought after than other abundant, more familiar, and "more tasty game, e.g. pigeons and parrots" (82).

3. The question of "cause" is complex. Of course, humans introduced pigs, rats, monkeys, and other animals to Mauritius—as well as ate a lot of Dodos themselves—but these introduced animals have their own agency that should not be denied in efforts to place the blame on humans (however rightly). We might also ask about the extent to which we ourselves inherit culpability for the actions of past generations (on this theme, in different contexts, see Bastian [2012b] and Clark [2007]). In short, there are various ways in which "our" implication in the extinction of others might be understood. This book is an effort to think through some of these possibilities.

4. As outlined in chapter 3, I think that the notion of "habitat loss," and thinking about animals as occupants of "habitats" more generally, is deeply problematic. I have used the term here because it is the one that most commonly appears in the literature.

5. "Background extinction" is the "normal" level of extinction expected as part of underlying evolutionary processes in which species are constantly coming into and going out of existence (discussed further in chap. 1). With this is mind, it might be thought that viewing current anthropogenic extinction as something distinct from "background extinction"

implies that it is somehow outside "normal evolutionary processes." While it might make sense to think about an asteroid-induced mass extinction in this way, the notion that the actions of one species (co-evolved with countless others on this planet) should be viewed as not a part of these processes is clearly conceptually problematic on some level. In terms of the scale of our impact on the diversity of living forms, however, at the present point in evolutionary history *Homo sapiens* clearly has more in common with an asteroid than it does with any other species. In reality, however, the definition of a "mass extinction" does not rest on the notion that the cause is external to normal evolutionary processes (that is, extraterrestrial in origin). Rather, it rests on the identification of a pattern of loss that is (1) temporally brief in terms of geologic terms, (2) broad in terms of the taxonomic diversity of the species affected, and (3) occurring at a much higher rate than that normally found in the fossil record (Raup and Sepkoski 1982).

6. On endangered animals and charisma, see Lorimer (2007); on the extinction of unknown species, see Smith (2011).

7. This "lively" approach to storytelling is the result of an ongoing collaboration with Deborah Bird Rose and Matthew Chrulew. For the past four years, together the three of us have been talking, thinking, and writing about extinctions and how it is that we might best tell extinction stories. At present, we are beginning work on a paper that takes up this topic in a more proscriptive matter. This discussion is also part of a wider collaboration with the Extinction Studies Working Group (www.extinctionstudies.org).

8. For further discussion of the extinction of the Passenger Pigeon, see Albus (2011) and Allen (2009).

1. FLEDGING ALBATROSSES

1. The "pelagic zone" is the area of the world's oceans that is far removed from land, either the coast or the seafloor.

2. In order to achieve this efficient form of flight, albatrosses alternate between periods of gliding in a steadily descending movement toward the ocean's surface and short, very abrupt periods of upward movement in which they position their long outstretched wings to catch the wind and momentarily lift them skyward. Even the seemingly energetic task of holding wings outstretched is made easy for the albatross by means of a tendon that locks the wings in place, requiring little muscular effort (Lindsey 2008:66; Safina 2007; Shaffer 2008:152).

3. Not all species of albatross commonly breed each year (some breed only in alternate years). This dynamic of returning to particular places to breed each year (side fidelity and philopatry) is discussed in greater detail in chapter 3, in the context of an urban colony of Little Penguins that makes its home along a disappearing shoreline in Sydney Harbour, Australia.

4. I take the term "long engagement" from Olsen and Joseph (2011; see also Lindsey 2008:83–84). Undoubtedly, the best-known aspect of albatross courtship is the elaborate dances performed by paired birds. Intricate combinations of bodily gesture and vocalization are synchronized by two birds, moving in response to each other, to produce what Lancelot E. Richdale called the "Ecstatic Ritual" (quoted in Rice and Kenyon 1962:530). It is impossible to capture in words the sound and sight of these courtship dances—wings outstretched, bodies bowed to each other, long necks reaching up to perform the "sky call." Even once the birds have settled into a pair, in future years when they return to the island to breed they will again sing and dance together for the period between arrival and the building of a nest. Albatrosses are largely monogamous, so pair-bonds usually endure until broken by the death or disappearance of one of the birds, with "divorces" being very rare (in this context, "divorce" refers to a situation in which both members of a breeding pair are known to still be alive, but are no longer breeding with each other [Rice and Kenyon 1962:524]). If a bond is broken, however, it usually takes more than a year for a new one to be established.

5. In practice, it is often not easy to determine what constitutes the "beginning" (or the "end," for that matter) of a species. There is an important distinction here between "speciation" and "phyletic evolution." The former refers to the splitting off of a group within an existing species, such that over time the new (reproductively isolated) group, responding to different selection pressures, heads down a distinct evolutionary path and a new species emerges. The latter, by contrast, is an ongoing process of change within a species (without splitting) that is significant enough to produce a species different from the one that previously existed (Mayr 2001:177).

It should also be noted that this distinction between Darwinian and pre-Darwinian understandings should not be overdrawn. As a growing body of literature in the history and philosophy of biology is now showing, pre-Darwinian notions of what constitutes a "biological species" were far from homogeneous (Amundson 2005; Wilkins 2009). As Ron Amundson (2005:36) makes clear, until roughly 100 years before Darwin, Western philosophy and science tended to regard "species" as relatively fluid. In particular, the popularity of "transmutationist" views encouraged a belief that plants or animals of one kind might readily change into those of another. These transformations could take place either within a single generation (through metamorphosis or subtler adaptations to climate) or across generations (through hybridization). It was through these avenues that a barnacle might become a goose, maize might be transformed into wheat, or a giraffe might come into being through the pairing of a camel and a leopard.

For obvious reasons, these transmutationist understandings played havoc with efforts to create a systematic taxonomy of living things. Through the work of Carl Linnaeus (1707–1778), his students, and others, however, empirical evidence began to emerge to indicate that these kinds of radical transformations did not occur and that species were actually fixed entities (Amundson 2005:34–41). John Wilkins (2009:95), like Amundson, argues that species fixism emerged much later in the history of naturalism, but he "credits" John Ray (1627–1705), not Linnaeus, with its invention.

However it arose, this understanding of species as relatively stable entities provided an important grounding for many of the taxonomic efforts of the period—especially the Natural System now associated with Linnaeus's work. In this historical context, species fixity represented an important scientific advance over previous transmutationist understandings. It was these fixist views, however, that the growing acceptance of evolutionary theory soon replaced.

Through the work of Darwin and others, from the middle of the nineteenth century, it was increasingly accepted that species are involved in ongoing processes of evolution (although the speed and mechanisms through which this evolutionary change occurs remained, and to some extent still remain, controversial). As a result, taxonomic efforts to organize the diversity of life gradually shifted from an ahistorical comparison of morphological features (type) to an expression of evolutionary history and phylogenetic relatedness. At the core of this transition is Darwin's notion of "common descent." Whereas previous proponents of evolution tended to assume a unique creation event for each species (which may then evolve over time), and so essentially singular phyletic lineages, "one of Darwin's major contributions was to have proposed the first consistent theory of *branching evolution*" (Mayr 2001:19). The Natural System becomes a family tree of sorts, a Tree of Life (Amundson 2005:133).

6. These characteristics of a species are inherited in the form of complex developmental systems that include genetic and various extra-genetic dimensions (Jablonka and Lamb 2005; Oyama 2000). For a slightly more detailed discussion of inheritance, see chapter 3. Of course, there is a great deal of individual variation within each species, variation that, as Darwin noted, is vital to the dynamic and evolving nature of life (Mayr 1996).

7. At Midway, this destructive human presence has taken many forms over the years. From the late nineteenth century, albatrosses in this area were killed in the hundreds of thousands by Japanese feather hunters who stripped breast and wing feathers and left the rest of the bird to rot. These feathers made their way around the world, as stuffing for bedding, but also to supply the growing demand of the fashion industry, especially for the adornment of hats (De Roy 2008:111–12). From the early twentieth century, the atoll's convenient midway location resulted in a series of ongoing disturbances: first as a way station for telegraphs between Asia and the United States; then as a refueling spot for early trans-Pacific flights; and finally, from the 1940s until the 1990s, as a major base for the U.S. Navy. During this time, the navy completely transformed the atoll, replacing breeding grounds with buildings, aircraft hangars, and runways. In this environment, albatrosses were both accidentally killed—ensnared in cables and antennas or through collisions with aircraft—and deliberately killed in the tens of thousands in an effort to reduce these collisions (De Roy 2008:113; Lindsey 2008:104–5). These threats have now mostly subsided, and since the 1990s the atoll has been a National Wildlife Refuge, administered by the U.S. Fish and Wildlife Service (Lindsey 2008:105).

8. This dubious honor excludes families with only one species. It should also be noted that over the past several decades, many fishing vessels, governments, and nongovernmental organizations around the world have developed and adopted a range of technologies or changed practices in an effort to reduce this mortality. Despite their considerable achievements, mortality levels remain very high and bycatch is still a very significant threat to the continuity of albatross species both in the North Pacific and around the world (Arata, Sievert, and Naughton 2009:23; Molloy, Bennett, and Schroder 2008; Sullivan 2008). For several decades, the high-seas squid and salmon drift-net fishery was also a central part of this spiraling albatross mortality in the North Pacific, until it was officially closed by a United Nations resolution in 1992 (Naughton, Romano, and Zimmerman 2007:10). Pelagic long-line fisheries, however, are still common in the region.

9. But, of course, it is not just albatrosses that are exposed to danger here: a range of other birds and mammals eat at the upper trophic levels, ourselves included. Striped dolphins in the North Pacific Ocean, for example, have PCB and DDT levels that are 13 million and 37 million times higher, respectively, than the concentrations found in the waters they inhabit (Thornton 2000:25). We are poisoning and contaminating our oceans: from the polar regions to the equator, from whales to the smallest bacteria; wherever we look closely, we find our presence in this most destructive and insidious of forms.

10. Among the albatrosses of Midway, this toxic burden has been borne primarily by the Black-foots. Likely as a result of their different foraging areas during the long breeding season, each species seems to be exposed to a different level of contamination (Finkelstein et al. 2006). On average, Black-foots have two to five times the levels of PCBs and DDTs that Laysans do (Finkelstein et al. 2006). Both species, however, possess levels of contamination that are one to two orders of magnitude higher than those of the albatrosses of the Southern Ocean (Guruge, Tanaka, and Tanabe 2001).

11. Kaua'i is at the northwest edge of the main chain of the Hawaiian Islands. Before the arrival of Polynesians on these islands about 1,500 to 2,000 years ago, albatrosses would have nested here in very large numbers. After being almost completely extirpated from this and other larger islands in the chain, in recent years a handful of birds have returned to establish small colonies—usually within fenced or otherwise protected areas where there is a reduced threat from recently introduced predators, such as domestic dogs. Today, there are around 200 pairs of nesting birds on Kaua'i, but due to the dedicated work of some locals, this number is growing. See, in particular, the work of the Kaua'i Albatross Network (www.albatrosskauai.org).

12. On the complex notion of "wild animals," see chapter 4.

13. For further discussion of the evolved "sensitivities" of birds to environmental change and the need to think about them outside of simple, hierarchical notions of "intelligence" (or its absence), see chapter 3.

14. This albatross experience took place with Hob Osterlund and Deborah Bird Rose. Osterlund runs the Kaua'i Albatross Network. When we visited these albatrosses together, she remarked on the peculiar "trust" that these birds show toward people and so initiated

the process of thought that led to some of the central strands of this chapter. I thank Osterlund for sharing her time and her insight on albatrosses, death, and a range of other topics. Thank you also to Michelle Bastian (pers. comm.) for her framing of this situation as that of inhabiting a "geological moment."

15. While it may well be possible to find meaning and value through a temporal frame locked on periods of millions of years, this is not the approach of this chapter (Rolston 1998). George Levine (2006) has also offered a convincing argument for the need to move away from understandings of evolutionary theory as a necessarily disenchanting discourse.

16. Of course, the kinds of time frames within which evolution takes place are very variable, depending on the organisms in question and a range of other factors. In this context, the slow, extended, temporal frames of evolution are always relative (Hird 2009; Oyama 2000:4).

17. This comment was made by James Hatley at the first meeting of the Extinction Studies Working Group on the southeast coast of Australia, February 13–17, 2012. For more information, see http://extinctionstudies.org/.

18. For example, roughly 39 percent of all Laysan Albatross pairs nesting on the island of O'ahu are female–female pairs, many of them engaged in long-term partnerships (Young, Zaun, and VanderWerf 2008).

19. As the recent controversy over the film *March of the Penguins* (2005) has demonstrated so well, we must remain wary of characterizations of avian (and other nonhuman animal) reproduction that dishonestly squeeze birds' lives into anthropomorphic, and often hetero-normative, frameworks. This is particularly worrying when, as in this case, "zoomorphic" reasoning is then used to extract lessons about "proper" or "natural" ways of life for human individuals and communities from (supposed) bird behavior (Wexler 2008). I am mindful that the reproductive labor of the albatrosses discussed in this chapter is, to some extent, readily "recognizable" and "relatable" precisely because of its surface similarity to the reproductive processes of (some) humans: from "nuclear family" units to the sacrifice, labor, and obvious care by parents. Although I am here thinking with albatrosses as a way into this space, my hope is that—in different ways, ways that will themselves require a great deal of additional fleshing out—a similar case may be made for other forms of speciated life.

20. In this context, a species is simultaneously an open and a closed "community" of beings. Reproductively, species are (ordinarily) closed off to others: through various "isolating mechanisms"—ecological, morphological, behavioral, genetic, or otherwise—successful interbreeding between organisms of different species is usually prevented (Mayr 2001:169–70). On one level, it is this "closure" that makes adaptation possible: the species (or its local population) providing a relatively isolated group among which traits may be selected for or against, and so proliferate or disappear. In this context, species emerge and are continually re-formed through their isolation. But this relative isolation of a species over vast temporal horizons is coupled with other important forms of long-term open-

ness in which species are entangled in diverse relationships of interaction, nourishment, co-evolution, and more. In short, it is an isolation that is shot through with connectivities of other kinds, and yet it is an important form of isolation nonetheless. In this context, the ongoing life of a species, like the other "biological systems" that Cary Wolfe (2009) discusses, is made possible through an "autopoietic closure, on the basis of which—and only on the basis of which—it can engage in various forms of 'structural coupling' "(xxii). In short, species take shape within a partial, and yet fundamental, form of isolation that is constitutive of the possibility of their ongoing flourishing as part of broader and more diverse communities of life.

21. How to inhabit a space in which no one can simply be dismissed as "ethically irrelevant" is a complex question, requiring situated ethics attentive to particularities. The remaining chapters of this book offer some thoughts in this direction.

22. On the complicated nature of care in multispecies contexts, see chapter 4.

23. On the evolution of affective engagements in relation to grief and empathy, see chapter 5.

24. Deborah Bird Rose's work here draws on conversations with the Australian Aboriginal people of the Victoria River Region in the Northern Territory.

25. I take the term "Cenozoic achievement" from Hatley (2012). This is an approach that, as Val Plumwood suggests we must, rejects the simplistic divide between "shallow" and "deep" environmental ethics. The task is not to choose between human well-being "us" and nonhuman well-being "them," but to find ways to cultivate and value mixed ecological communities that include humans (Plumwood 2009:116).

2. CIRCLING VULTURES

1. This understanding of species as "flight ways" is outlined in detail in chapter 1.

2. All references to "Houston, pers. comm.," refer to David C. Houston, e-mail exchange with author, mid-2009. Houston, Honorary Senior Research Fellow at the University of Glasgow, is one of the world's foremost experts on vulture biology and behavior.

3. While vultures sometimes eat "badly decomposed" food, they do prefer food that is relatively fresh (Houston 2001).

4. There are numerous other dimensions of the lives of cattle in India that cannot be understood as anything other than tragic. While almost all Indian states ban the slaughter of cattle, in many cases this has simply meant that slaughter is carried out illegally—and thus in a completely unregulated manner—or that cattle are subjected to long and crowded transportation to slaughterhouses in neighboring states or countries (Singh 2003).

5. All references to "Cunningham, pers. comm.," refer to Andrew Cunningham, interview with author, London, September 11, 2008. Cunningham, a wildlife epidemiologist at the Institute of Zoology, Zoological Society of London, works on the conservation of vultures in India.

2. *Circling Vultures*

6. Here, and in what follows, I have drawn on statistics in an effort to convey the immensity and the inequity of the problems developing in India in the absence of vultures. I am mindful, however, that numbers cannot really do this and that statistics not only fail to capture suffering, but also can undermine the ethical demands that it issues (van Dooren 2010). Despite these drawbacks, in this case the numbers seem to tell a vital part of the story.

7. For a very different discussion of environmental justice and direct relationships with places/ecologies, see Plumwood (2008a).

8. The ecological notion of "functional extinction" provides a terminology to describe a species that is extant, but in such reduced numbers that it no longer fulfills its previous ecological roles. This concept is in some ways helpful, but it cannot do all the work that the "dull edge" might. Functional extinction still fundamentally ties "extinction proper" to the death of the last of a kind, and the other losses that it draws our attention to are (understandably, given the disciplinary origins of the term) purely "ecological."

3. URBAN PENGUINS

1. Perhaps this is what ultimately made the wall "remarkable" in its destructive potential, its being among the last in a long sequence of walls.

2. The first wall that I have seen documentary evidence for appears on a subdivision map of 1914, drawn up when the original property was broken into fourteen lots to be sold off. Over the past century, each of these lots has been subdivided many more times.

3. I have borrowed the terminology of "lost places" from Peter Read's moving study *Returning to Nothing: The Meaning of Lost Places* (1996).

4. Many of the ideas in this chapter have their origin in a collaboration with Deborah Bird Rose. See, for example, van Dooren and Rose (2012).

5. I will set aside the ongoing discussion about whether space precedes place in any meaningful way (as has often been assumed) or, instead, lived-place should be understood as prior and space as an abstraction from it (Casey 1996). It is not necessary to resolve this debate in order to appreciate that "place" might be understood as an embodied, lived, and meaningful environment.

6. I am inclined to think of this not as a linear spectrum, but as a "diversity of sensitivities," as discussed later in the chapter. But Darwin's basic point about continuity remains.

7. All references to "Challies, pers. comm.," refer to Chris Challies, correspondence with author. Challis is a recognized expert on Little Penguin philopatry, fidelity, and breeding behavior (his work is actually focused on a Little Penguin subspecies: the White-flippered Penguin [*Eudyptula minor albosignata*]).

8. Another study, focusing on the Manly colony's closest penguin neighbors—at Lion Island—found a high degree of nest fidelity, but did not detect any significant relationship between breeding success and nest changes (Rogers and Knight 2006).

9. Most of these explanations for fidelity are rooted in economic notions of "competitive advantage." More specifically, the benefits conferred by site fidelity are thought to improve penguins' breeding success. While this may well be the case, we should be careful about allowing these evolutionary explanations to become exhaustive accounts of animal behavior in a way that either negates or obviates richer notions of nonhuman cognitive life (Crist 1999). In short, all these practical advantages of fidelity—which make good evolutionary sense—tell us nothing about what site and mate fidelity *feel* like to Little Penguins: how the imperative to be reunited with a place or a partner is experienced by individuals and comes to animate understandings, actions, and relationships. On the important difference between function and motivation, see de Waal (2008).

10. I discuss this casting of penguins as "unwanted guests" in more detail later in the chapter.

11. Beyond the animals, like penguins, that possess more readily recognizable forms of meaningful relationships with their environments, a whole host of organisms, including plants and bacteria, "trade in signs and wonders" (Haraway 1997:8) in ways that deserve recognition. They, too, are part of a broader understanding of "nature in the active voice" (Plumwood 2009). For the sake of simplicity, however, I have confined my discussion in this chapter to penguins (and a few other species with perhaps more recognizable, from a human perspective, ways of making meaning).

12. This inheritance is definitively more than genetic. Contrary to the stories that we often like to tell about evolution and inheritance, much more is gifted between generations of nonhuman animals than a genotype. Through diverse processes of social learning and exposure to particular experiences and environments in early life, many animals inherit traits, behaviors, languages, skills, and other "cultural traditions" that they will, in turn, pass on to their own offspring (Jablonka and Lamb 2005; Oyama 2000).

4. BREEDING CRANES

1. References to "Duff, pers. comm.," refer to Joe Duff, interview with author, June 25, 2012. Duff is the lead ultralight aircraft pilot for Operation Migration.

2. The ultralight aircraft–led migration of Canada Geese was popularized in the movie *Fly Away Home* (Ballard 1996).

3. At roughly the same time, Kent Clegg, James Lewis, and David Ellis (1997) were conducting similar trial migrations with Sandhill Cranes from Idaho to New Mexico. I have used the term "Whoopers" occasionally to refer to Whooping Cranes. This name is widely used in preference to Whooping Crane by conservationists and others who work with the birds on a daily basis.

4. All references to "French, pers. comm.," refer to John French, interview with author, June 30, 2012. French, a wildlife biologist, is the manager of the Whooping Crane project at the Patuxent Wildlife Research Center, Laurel, Maryland.

5. In recent years, however, some Whoopers at ICF have been given Sandhill Crane eggs with which to "practice" incubating and rearing. In addition, there have been some small trials of parent rearing (Tarr, pers. comm.). All references to "Tarr, pers. comm.," refer to Bryant Tarr, interview with author, June 27, 2012. Tarr is the curator of birds at ICF, Baraboo, Wisconsin.

6. Later work on imprinting by Eckhard Hess (1964) and others has established that this following behavior itself may play an important role in cementing a chick's attachment to its parental object.

7. Things are very different for those birds that will never be released, but remain within the captive breeding population for the entirety of their lives. The problems and possibilities that imprinting and habituation raise for these birds are discussed in more detail later in this chapter.

8. Contrary to some popular statements, imprinting has nothing to do with a lack of intelligence. According to an article in the *Guardian*, Tony Whitehead, public-affairs officer of the Royal Society for the Protection of Birds, recently commented, "Birds have small brains[;] with basic disguising, the young cranes are perfectly happy to accept rigid wings, engines and even [people] as mummy and daddy" (quoted in Malein 2012). Even birds as intelligent (from a human perspective) as corvids—which are very intelligent indeed (chap. 5)—imprint on their parents. In short, this developmental process, like the strong philopatry of Little Penguins (chap. 3), must be understood as an evolved and specific form of sensitivity, not as a black-and-white measure of intelligence or a lack thereof.

9. But for organisms whose species is so close to the edge of extinction, is this necessarily a good, or sustainable, option? The sad reality is that it may well not be. Perhaps being something other than a Whooping Crane—even if "only" behaviorally and socially—is a better option for these birds. This is a question that other captive breeding programs have also needed to take up, asking whether captive animals might be trained to live differently once released in order to increase their chances of survival—perhaps through alternative foraging or nesting strategies, or learning to exploit new food sources (see, for example, the short discussion about this possibility for the Hawaiian Crow in chap. 5). It is highly debatable, however, that cross-species imprinting might be a valuable part of these kinds of efforts for released birds. Later in this chapter, I consider the possibility of cross-species imprinting for birds that will spend the entirety of their lives in captivity as part of breeding populations.

10. I suspect that Vinciane Despret herself would be critical of the ethics of some of Konrad Lorenz's relationships with birds and other animals. My discussion in this chapter has focused on a single paper from Despret's (2004a) substantial body of work (much of which is, unfortunately, unavailable in English). My colleague Jeff Bussolini (2013), an attentive reader of Despret's work in French, informs me that much of her work explores issues of politeness and sensitivity in interspecies relationships that are aligned with my own position. And so while I am not suggesting that Despret is endorsing the ethics of Lorenz's relationships with imprinted birds, in this particular article her emphasis on the innovative

and caring possibilities in his approach to research overshadows some of the violences of the broader context within which his experiments took place.

11. The fascinating short film "My Life as a Turkey" (Allen 2011) provides an example of the kind of 24/7 dedication that might be required to rear human-imprinted birds in a way that attempts to put their needs first. Ultimately, however, it is an ambiguous story, full of complex and fraught relations.

12. This is not about keeping or making birds "wild" in any simple sense. Rather, it seems that the categories of "wild" and "tame" are of little help here. As Clare Palmer's (2010) analysis of conceptions of wild animals shows so well, "the wild" is most often shorthand for "non-humanized" (whether this humanization is taken to occur though relocation to human-dominated landscapes, individual taming, or longer-term and more generalized relationships of "domestication"). This is a highly dualized framework in which the human is the measure of all things; an animal's way of being in the world is gauged solely in reference to its level of entanglement in human lives and projects. But these are not the only relevant cross-species relationships. Being imprinted on a Sandhill Crane, for example, is just as problematic for a Whooping Crane as being imprinted on a human. What is at issue here is not how "tame" or "wild" a Whooping Crane is, but the extent to which the broad social world in which each crane is enfolded is conducive to a flourishing life.

13. Of course, the type is "exhaustible" in an important sense, and it is against this possibility—in some cases *probability*—that these conservation projects work.

14. It is important to note in this context that there are not just two regimes of care at play here—although I am particularly interested in the intersection of two general sets of priorities. Within each of them, however, there is a great deal of variability. *How to care* for a species, or for any given individual, is itself a shifting and contested issue.

15. I have focused in these comments on the other *birds* drawn into this conservation project. Numerous other species might also be mentioned. As in many captive breeding programs, the animals that are bred specifically to be fed to an endangered species are also a significant sacrificial population (Bekoff 2010). In addition, the release of captive-bred animals is often accompanied by "habitat modifications" that have an impact on other lives in various ways. The most obvious of the negative impacts are felt by those animals—usually predators, but sometimes also competitors—that are culled (killed) to give released animals a better chance of survival. This has not been a major component of Whooping Crane reintroduction efforts; instead, for example, in central Florida bobcats have been trapped and relocated (Hughes 2008:145)—a process that can have its own negative impacts. Increasingly, however, it is being recognized that "predator control" is often a vital component of successful release or relocation programs (Fischer and Lindenmayer 2000).

16. The environmental philosopher Holmes Rolston III (1999) offers a typical example in his discussion of a goat "eradication" program on San Clemente Island, off the coast of California. Around 14,000 goats were shot (and many others trapped and removed) to conserve three endangered plant species. This action was justified in Rolston's view because the goats are not endangered and therefore are "replaceable"—as well as not being

"native" to the island. I am in wholehearted agreement with Rolston that the endangered status of the plants is ethically relevant (chap. 1). His discussion takes place in the context of a response to Peter Singer's utilitarian ethic, which Rolston charges with paying inadequate respect to nonanimal forms of life, as well as species. But like those of many other conservationists and environmental philosophers, Rolston's position has the opposite problem. The surety with which he declares the "right" thing to do, his confidence in the fact that conserving endangered species "outweighs" and so justifies the suffering and deaths of individuals, is, I think, deeply problematic. What are the limits of this "trumping" capacity of endangered species? How many goats can be killed to save a species of plant, and in what ways can they be killed? How much can they be required to suffer? Ongoing practices of "invasive species management" in the United States, Australia, and a number of other countries indicate that our tolerance for mass death and suffering in the name of conservation can be very high indeed (Rose 2008; van Dooren 2011a). The other side of this equation, as the Whooping Crane case makes clear, is the suffering of those individuals that are "made to live" in zoos and captive breeding facilities for the sake of the continuity of their own species (Chrulew 2011a). For how long, and in what conditions, can these beings be made to live?

17. As previously noted, this already happens with a small number of Whooping Crane young at ICF.

18. The "crane cam" can be accessed on the Web site of Operation Migration (http://www.operationmigration.org/crane-cam.html).

5. MOURNING CROWS

1. All references to "Lieberman, pers. comm.," refer to Alan Lieberman, interview with author, November 29, 2010. Lieberman is the director of Regional Conservation Programs at the Institute for Conservation Research, San Diego Zoo, and has had a long-term involvement in the conservation of birds in Hawai'i, including the Hawaiian Crow.

2. On threats to the Hawaiian Crow, see Banko, Ball, and Banko (2002) and USFWS (2009).

3. It should also perhaps be noted that while for many crow species it is precisely these characteristics that have allowed them to thrive in close contact with the constantly changing conditions of human societies—intelligence and socially learned behaviors enabling high levels of adaptation—for many island crows, exposed to a very different set of selection pressures, this has not been the case. All around the world, this has created a similar situation. In John Marzluff's (pers. comm.) words: "On almost every island, whether you're in the Caribbean or in the Pacific . . . there is a native crow . . . and it's almost always a frugivore, and it's almost always endangered." All references to "Marzluff, pers. comm.," refer to John Marzluff, telephone interview with author, November 13, 2010. Marzluff is a professor in the School of Forest Resources at the University of Washington. Much of his

research focuses on corvid ecology, behavior, and conservation, and he is a former member of the USFWS's 'Alalā Recovery Team.

There is little doubt that Hawaiian Crows possess the intelligence and adaptability to live alongside people, scavenging waste and taking advantage of our presence in a range of other ways. In fact, the 'Alalā Recovery Team discussed, and ruled out, the possibility of conserving the species by helping it to more quickly learn to utilize human rubbish. This option was dismissed not because it would be difficult—in fact, it would likely be a far simpler means of conserving the crows—but because the team viewed its role as breeding and releasing birds who would be "as wild, and frugivorous, and forest loving, as possible" (Marzluff, pers. comm.). This situation raises a range of interesting questions about the goals of conservation. In particular, what is to be conserved? Is it more than the biological and genetic diversity of a species? Is its behavioral (and perhaps cultural) repertoire also something worthy of preservation? If so, on what terms? These questions are in the background, and occasionally the foreground, of this chapter.

4. I will primarily refer to Martin Heidegger's *Dasein* (literally, "being there") as "the human." While this description is clearly overly simplistic, in the context of this discussion about Heidegger's distinction between the human and the animal, it is *Dasein*'s being as a human being that is its most salient feature.

5. A more recent example of this kind of view is offered by philosopher Jeff Malpas in his argument that the knowledge of one's own death is essential to the possession of a "life." As Malpas (1998) succinctly puts it: "To be a creature that has a life, to be a creature that has a world, to be a creature that has a sense of value and significance, is also to be a creature that has a grasp of the possibility of its own ending" (134). Although Malpas is not explicit about who does and does not know death, and so who does and does not have a life, his thinking clearly draws on a long tradition in which the salient distinction is that between the human and the animal. The only example that Malpas (1998:120–21) provides of a creature that does not have a life, and so presumably also has no knowledge of its own finitude, is his cat. His paper closes with a long quote from Heidegger's work, focused on the distinction between "man," who dies, and the animal, who merely perishes, and a final footnote that states that the aim of the paper has been to "develop an argument for what is essentially a Heideggarian conclusion, but without reliance on explicitly Heideggarian premises" (Malpas 1998:134). While Malpas perhaps leaves things a little more open than does Heidegger, it seems fair to say that his thought remains centered on an indefensible and assumed human–animal divide.

We see in both Heiddeger's and Malpas's work one of the central reasons why knowledge of death matters. Put simply, there is a notion here that in the absence of death one does not live fully. Malpas (1998) articulates this connection clearly in his opening sentences, drawing on Bernard Williams to argue that life in the absence of death would be "devoid of interest, devoid of meaning" (120). But as Jonathan Strauss (2000) has noted, this idea is by no means limited to these few thinkers, being found, for example, in the thought of Paul Tillich, Herman Feifel, Gillian Rose, and others. In Feifel's words: "[T]he notion of the uniqueness and individuality of each one of us gathers full meaning only in realizing that we are finite" (quoted in Strauss 2000:93). In this context, as Strauss (2000) goes on to note, humans must—in order to assume our individuality—"sacrifice that animal existence in ourselves, kill off a somnolent living without a life, an existence without death or self" (101).

6. I do not know the answer to this question, but it is clear that any reasonable speculation must engage with the current ethological literature on the topic. In marked contrast, it is interesting to note how infrequently most philosophers reference this literature at all before making their pronouncements. See also Calarco (2008).

7. As these behaviors and expressions—including, for example, birdsong—are taught and travel through generations (chaps. 1 and 3), they must carry the insights and developments of the dead in one way or another, shaping the lives of the current generation. While there may well be differences between human and nonhuman interactions and inheritances in this context—and Françoise Dastur's (1996) precise position is somewhat unclear—it seems to me important to find ways to acknowledge the many kinds of inheritance and intergenerational continuity that occur among nonhumans, and not to jump into the acceptance of another dualistic and anthropocentric "proper."

8. The context for this assertion with reference to the Hawaiian Crow should be clear from this chapter and other work cited in it. For a discussion of human violence toward elephants, see Poole (1996) and Wylie (2010). On elephants' stress and social breakdown, see Bradshaw (2004) and Bradshaw et al. (2005). This situation offers an important example of the mutually reinforcing logic that Matthew Chrulew (2011b) has noted in relation to philosophies of animals. He points to the relationship between captive animals in zoos and elsewhere and the kind of philosophical thinking that represents animals as fundamentally "captive" (for example, Heidegger). Here, ideas about animals as lesser subjects are formed through interactions with, or are justified with reference to, animals that we increasingly force to live in diminished conditions with limited and disturbed socialities.

9. I have used the terms "grief" and "mourning" interchangeably in this chapter, in opposition to more conventional usage, which often reserves the latter word exclusively for humans. I suspect that this conventional usage stems from the fact that "grief" is often employed to refer to responses to loss in general, whereas "mourning" is specifically a response to the loss brought about by death (Attig 1996:9). If animals are unable to understand death, however, then they are unable to experience this specific kind of loss, and so unable to mourn properly—they are limited to grieving, as they would for any other lost attachment. As this chapter makes clear, I am not confident that things are this simple.

10. On anecdote as a respectable part of ethological study, see Bekoff (2007) and Crist (1999). In addition, it must be kept in mind that there are clear ethical problems with the construction of formal experiments to test for an emotion like grief.

11. These conclusions are given further weight by recent comparative work in neurology that highlights important similarities in neural circuits and neurotransmitters between

birds and mammals. It seems that the neural bases that enable grief in many mammals, including ourselves, are also found in the remarkable corvids (Marzluff 2012).

12. This is a view that Darwin seems to have shared in a letter to his recently bereaved cousin: "Strong affections have always appeared to me, the most noble part of a man's character and the absence of them an irreparable failure; you ought to console yourself with thinking that your grief is the necessary price for having been born with (for I am convinced they are not to be acquired) such feelings" (quoted in Archer 1999:75).

13. For a discussion of the possibility of "culture" among New Caledonian Crows, see Hunt (1996) and a reply from Boesch (1996). W. C. McGrew (1998) also offers useful insight into various ways of understanding "culture" and its presence among nonhumans, with particular reference to nonhuman primates.

14. For a discussion of similar gatherings at sites of death by another corvid species, the Western Scrub-Jay (*Aphelocoma californica*), see Iglesias, McElreatha, and Patricelli (2012).

15. On the difference between "function" and "motivation," see de Waal (2008).

16. The "biological" and the "social" are all mixed up here in ways that undermine the coherence of any concrete distinction between them. Sociality in all its multiple forms is rooted in specific biological capacities—in this case, capacities that we might label "emotional" or "cognitive." In a related but distinct vein, plants and various microorganisms are also engaged in ongoing "social" relationships of their own kind—exchanging signs and meanings, communicating in ways that we often underestimate (Hall 2011). In this sense, sociality is perhaps a common feature of all life and should not be restricted to those organisms who possess modes of interaction similar enough to those of humans to be immediately recognizable as such (Hird 2009). In other words, our being social creatures, as well as the specific forms that this sociality takes, are in important ways features of our biological makeup. At the same time, however, biology has itself evolved within the context of very material processes of intergenerational life in the company of others. Sociable life produced the conditions for the evolution of various social capabilities, which, in turn, deepened and enhanced those social relationships. There is no sociality outside of its specific biological possibilities, nor is there any biological form that has not been shaped by its own particular social milieu.

17. It is with this understanding in mind that I would like to suggest, in contrast to some of Judith Butler's (2004, 2009) recent work on the topic, that mourning is less about the "recognition" of a valuable or "grievable" life, and more about the simple embodied reality of our being more or less affected by others, more or less constituted by their presence, more or less emotionally and intellectually bound up in their fate. It is this differential entanglement of affect that gives rise to the varied degrees of grief that accompany death and loss; we are simply more attached to and invested in the lives of some people, some animals, some environments, some jobs and belonging than we are in others. Not to mourn for the passing of some, or to mourn less for some than others, does not *necessarily* indicate

a failure to recognize a life as "grievable" or "worth living," as Butler suggests—although I accept that in some limited cases, it may mean precisely this. In most cases, however, the kind of threshold that Butler's terms imply—valuable or not, grievable or not—cannot do justice to the full spectrum of emotional responses that loss elicits, or fails to elicit, in the countless lives of all those variously connected and entangled others that are left behind. In contrast, I would suggest that a "failure" to grieve may more often result from an inability to "get" (at various experiential levels) how one's own life and world are shared with these dying others.

18. Thanks to Michelle Bastian for pointing out this connection to Matei Candea's (2010) wonderful work on meerkats.

19. For a detailed overview of avian extinctions in the Pacific region, see Steadman (2006).

REFERENCES

Agence France-Presse. 2007. "Nepali Vulture 'Restaurant' Aims to Revive Decimated Population." Agence France-Presse, October 29. http://afp.google.com/article/ALeqM5hXfBY76myTDIuT9ACEa0mIBcoRVw (accessed July 2, 2013).

Aitken, G. M. 1998. "Extinction." *Biology and Philosophy* 13:393–411.

Albus, Anita. 2011. *On Rare Birds*. Translated by Gerald Chapple. Sydney: New South.

Allen, Barbara. 2009. *Pigeon*. London: Reaktion Books.

Allen, Colin, and Marc Bekoff. 1999. *Species of Mind: The Philosophy and Biology of Cognitive Ethology*. Cambridge, Mass.: MIT Press.

Allen, David. 2011. "My Life as a Turkey." *Nature*, PBS, November 16.

Amadon, Dean. 1983. Foreword. In *Vulture Biology and Management*, edited by Sanford R. Wilbur and Jerome A. Jackson. Berkeley: University of California Press.

Amundson, Ron. 2005. *The Changing Role of the Embryo in Evolutionary Thought: Roots of Evo-Devo*. New York: Cambridge University Press.

Angst, Delphine, Eric Buffetaut, and Anick Abourachid. 2011a. "In Defence of the Slim Dodo: A Reply to Louchart and Mourer-Chauvire." *Naturwissenschaften* 98:359–60.

———. 2011b. "The End of the Fat Dodo? A New Mass Estimate for *Raphus cucullatus*." *Naturwissenschaften* 98:233–36.

APCRI (Association for the Prevention and Control of Rabies in India). 2004. *Assessing the Burden of Rabies In India: WHO Sponsored National Multi-centric Rabies Survey 2003*. Bangalore: Association for the Prevention and Control of Rabies in India.

Arata, Javier A., Paul R. Sievert, and Maura B Naughton. 2009. *Status Assessment of Laysan and Black-footed Albatrosses, North Pacific Ocean, 1923–2005*. Scientific Investigations Report 2009–5131. Reston, Va.: U.S. Geological Survey.

Archer, John. 1999. *The Nature of Grief: The Evolution and Psychology of Reactions to Loss*. London: Routledge.

Attig, Thomas. 1996. *How We Grieve: Relearning the World*. New York: Oxford University Press.

Auman, Heidi J., James P. Ludwig, John P. Giesy, and Theo Colborn. 1997. "Plastic Ingestion by Laysan Albatross Chicks on Sand Island, Midway Atoll, in 1994 and 1995." In *Albatross Biology and Conservation*, edited by Graham Robertson and Rosemary Gales. Chipping Norton, Australia: Surrey Beatty.

Auman, Heidi J., James P. Ludwig, Cheryl L. Summer, David A. Verbrugge, Kenneth L. Froese, Theo Colborn, and John P. Giesy. 1997. "PCBS, DDE, DDT, and TCDD-EQ in Two Species of Albatross on Sand Island, Midway Atoll, North Pacific Ocean." *Environmental Toxicology and Chemistry* 16, no. 3:498–504.

Australian Women's Weekly. 1956. "When Summer Comes . . . Penguins at the Bottom of Their Garden." December 12, 22–23.

Ballard, Carroll. 1996. *Fly Away Home*. Culver City, Calif.: Columbia Pictures Corporation.

Banko, Paul C., Donna L. Ball, and Winston E. Banko. 2002. "Hawaiian Crow (*Corvus hawaiiensis*)." In *The Birds of North America Online*, edited by Alan Poole. Ithaca, N.Y.: Cornell Lab of Ornithology. http://bna.birds.cornell.edu/bna/ (accessed August 8, 2013).

Barad, Karen. 2007. *Meeting the Universe Halfway: Quantum Physics and the Entanglement of Matter and Meaning*. Durham, N.C.: Duke University Press.

Baral, Nabin, Ramji Gautam, Nilesh Timilsina, and Mahadev G. Bhat. 2007. "Conservation Implications of Contingent Valuation of Critically Endangered White-rumped Vulture *Gyps bengalensis* in South Asia." *International Journal of Biodiversity Science and Management* 3:145–56.

Barlow, Connie. 2000. *The Ghosts of Evolution: Nonsensical Fruit, Missing Partners, and Other Ecological Anachronisms*. New York: Basic Books.

Barnes, David K. A., Francois Galgani, Richard C. Thompson, and Morton Barlaz. 2009. "Accumulation and Fragmentation of Plastic Debris in Global Environments." *Philosophical Transactions of the Royal Society B: Biological Sciences* 364:1985–98.

Barnosky, Anthony D., Nicholas Matzke, Susumu Tomiya, Guinevere O. U. Wogan, Brian Swartz, Tiago B. Quental, Charles Marshall, Jenny L. McGuire, Emily L. Lindsey, Kaitlin C. Maguire, Ben Mersey, and Elizabeth A. Ferrer. 2011. "Has the Earth's Sixth Mass Extinction Already Arrived?" *Nature*, March 2, 51–57.

Bastian, Michelle. 2011. "The Contradictory Simultaneity of Being with Others: Exploring Concepts of Time and Community in the Work of Gloria Anzaldua." *Feminist Review* 97:151–67.

———. 2012. "Fatally Confused: Telling the Time in the Midst of Ecological Crises." *Environmental Philosophy* 9, no. 1:23–48.

———. 2013. "Political Apologies and the Question of a 'Shared Time' in the Australian Context." *Theory, Culture & Society* 30, no. 5: 94–121.

Bataille, Georges. 1997. "Death." In *The Bataille Reader*, edited by Fred Botting and Scott Wilson. Oxford: Blackwell.

BBC. 2011. "Japan Tsunami: Thousands of Seabirds Killed Near Hawaii." BBC News, Asia-Pacific, March 16. http://www.bbc.co.uk/news/12756033 (accessed July 2, 2013).

Bekoff, Marc. 2006. "Animal Passions and Beastly Virtues: Cognitive Ethology as the Unifying Science for Understanding the Subjective, Emotional, Empathic, and Moral Lives of Animals." *Human Ecology Forum* 13, no. 1:39–59.

———. 2007. *The Emotional Lives of Animals*. Novato, Calif.: New World Library

———. 2010. "First Do No Harm." *New Scientist*, August 28, 24–25.

———. 2012. " 'Zoothanasia' Is Not Euthanasia: Words Matter." *Psychology Today*, August 9. http://www.psychologytoday.com/blog/animal-emotions/201208/zoothanasia-is-not-euthanasia-words-matter (accessed July 2, 2013).

Birdlife International. 2008. *State of the World's Birds*. Cambridge: Birdlife International.

Blanco, J. M., D. E. Wildt, U. Hofe, W. Voelker, and A. M. Donoghue. 2009. "Implementing Artificial Insemination as an Effective Tool for Ex Situ Conservation of Endangered Avian Species." *Theriogenology* 71:200–213.

Boesch, Christophe. 1996. "The Question of Culture." *Nature*, January 18, 207–8.

Bourne, Julie, and Nicholas I. Klomp. 2004. "Ecology and Management of the Little Penguin *Eudyptula minor* in Sydney Harbour." In *Urban Wildlife: More Than Meets the Eye*, edited by Daniel Lunney and Shelley Burgin. Mosman, Australia: Royal Zoological Society of New South Wales.

Bowkett, Andrew E. 2009. "Recent Captive-Breeding Proposals and the Return of the Ark Concept to Global Species Conservation." *Conservation Biology* 23, no. 3:773–76.

Bradshaw, G. A., Allan N. Schore, Janine L. Brown, Joyce H. Poole, and Cynthia J. Moss. 2005. "Social Trauma: Early Disruption of Attachment Can Affect the Physiology, Behaviour and Culture of Animals and Humans over Generations." *Nature*, February 24, 807.

Bradshaw, Isabel Gay A. 2004. "Not by Bread Alone: Symbolic Loss, Trauma, and Recovery in Elephant Communities." *Society and Animals* 12, no. 2:143–58.

Brault, Pascale-Anne, and Michael Naas. 2001. "Editors' Introduction: To Reckon with the Dead: Jacques Derrida's Politics of Mourning." In *The Work of Mourning*, by Jacques Derrida. Edited by Pascale-Anne Brault and Michael Naas. Chicago: University of Chicago Press.

British Medical Association. 1995. *The BMA Guide to Rabies*. Oxford: Radcliffe Medical Press, on behalf of the British Medical Association.

Buchanan, Brett. 2008. *Onto-Ethologies: The Animal Environments of Uexküll, Heidegger, Merleau-Ponty, and Deleuze*. Albany: State University of New York Press.

Bugnyar, Thomas. 2011. "Knower–Guesser Differentiation in Ravens: Others' Viewpoints Matter." *Proceedings of the Royal Society B: Biological Sciences* 278, no. 1705:634–40.

Bugnyar, Thomas, and Bernd Heinrich. 2006. "Pilfering Ravens, *Corvus corax*, Adjust Their Behaviour to Social Context and Identity of Competitors." *Animal Cognition* 9:369–76.

Bull, Leigh. 2000. "Fidelity and Breeding Success of the Blue Penguin *Eudyptula minor* on Matiu-Somes Island, Wellington, New Zealand." *New Zealand Journal of Zoology* 27:291–98.

Bussolini, Jeffrey. 2013. "Recent French, Belgian and Italian Work in the Cognitive Science of Animals: Dominique Lestel, Vinciane Despret, Roberto Marchesini and Giorgio Celli." *Social Science Information* 52, no. 2:187–209.

Butler, Judith. 2004. *Precarious Life: The Powers of Mourning and Violence*. London: Verso.

———. 2009. *Frames of War: When Is Life Grievable?* London: Verso.

Butvill, Dave Brian. 2004. "Dances with Cranes." *California Wild*, spring.

Byrne, Denis, Heather Goodall, and Allison Cadzow. 2013. *Place-making in National Parks: Ways That Australians of Arabic and Vietnamese Background Perceive and Use the Parklands Along the Georges River, NSW*. Sydney: Office of Environment and Heritage.

Cade, Tom J. 1988. "Using Science and Technology to Reestablish Species Lost in Nature." In *Biodiversity*, edited by E. O. Wilson. Washington, D.C.: National Academy Press.

Calarco, Matthew. 2002. "On the Borders of Language and Death: Derrida and the Question of the Animal." *Angelaki: Journal of the Theoretical Humanities* 7, no. 2:17–25.

———. 2008. *Zoographies: The Question of the Animal from Heidegger to Derrida*. New York: Columbia University Press.

Candea, Matei. 2010. "I Fell in Love with Carlos the Meerkat: Engagement and Detachment in Human–Animal Relations." *American Ethnologist* 37, no. 2:241–58.

Casey, Edward S. 1996. "How to Get from Space to Place in a Fairly Short Stretch of Time: Phenomenological Prolegomena." In *Senses of Place*, edited by Steven Feld and Keith H. Basso. Santa Fe, N.M.: School of American Research Press.

———. 2001. "J. E. Malpas's Place and Experience: A Philosophical Topography, Converging and Diverging in/on Place." *Philosophy & Geography* 4, no. 2:225–31.

Chapman, M. G., and F. Bulleri. 2003. "Intertidal Seawalls—New Features of Landscape in Intertidal Environments." *Landscape and Urban Planning* 62:159–72.

Chrulew, Matthew. 2011a. "Managing Love and Death at the Zoo: The Biopolitics of Endangered Species Preservation." In "Unloved Others: Death of the Disregarded in the Time of Extinctions," edited by Deborah Bird Rose and Thom van Dooren, special issue, *Australian Humanities Review* 50:137–57.

———. 2011b. "Reflections in Philosophical Ethology." Paper presented at the workshop "The History, Philosophy and Future of Ethology," Macquarie University, Sydney, February 19–21.

Clark, Nigel. 2007. "Living Through the Tsunami: Vulnerability and Generosity on a Volatile Earth." *Geoforum* 38:1127–39.

Clegg, Kent R., James C. Lewis, and David H. Ellis. 1997. "Use of Ultralight Aircraft for Introducing Migratory Crane Populations." In *Proceedings of the Seventh North American Crane Workshop*, edited by Richard P. Urbanek and Dale W. Stahlecker. Grand Island, Neb.: North American Crane Working Group.

Clewell, Tammy. 2009. *Mourning, Modernism, Postmodernism*. New York: Palgrave Macmillan.

Cooper, Alan, and David Penny. 1997. "Mass Survival of Birds Across the Cretaceous–Tertiary Boundary: Molecular Evidence." *Science* 275:1109–13.

Crist, Eileen. 1999. *Images of Animals: Anthropomorphism and Animal Mind*. Philadelphia: Temple University Press.

Cronon, William. 1992. "A Place for Stories: Nature, History, and Narrative." *Journal of American History*, March, 1347–76.

Culliney, Susan Moana. 2011. "I. Seed Dispersal by the Critically Endangered Alala (*Corvus hawaiiensis*). II. Integrating Community Values into Alala (*Corvus hawaiiensis*) Recovery." M.S. thesis, Colorado State University.

Curby, Pauline. 2001. *Seven Miles from Sydney: A History of Manly*. Manly, Australia: Manly Council.

Cuthbert, Richard, Jemima Parry-Jones, Rhys E. Green, and Deborah J. Pain. 2007. "NSAIDs and Scavenging Birds: Potential Impacts Beyond Asia's Critically Endangered Vultures." *Biology Letters* 3:91–94.

Daniel, T. A., A. Chiaradia, M. Logan, G. P. Quinn, and R. D. Reina. 2007. "Synchronized Group Association in Little Penguins, *Eudyptula minor*." *Animal Behaviour* 74:1241–48.

Darwin, Charles. 1871. *The Descent of Man and Selection in Relation to Sex*. Vol. 1. New York: Appleton.

———. (1859) 1959. *The Origin of Species*. Philadelphia: University of Pennsylvania Press.

———. (1872) 1965. *The Expression of the Emotions in Man and Animal*. Chicago: Phoenix Books.

Dastur, Françoise. 1996. *Death: An Essay on Finitude*. Translated by John Llewelyn. London: Athlone Press.

Davis, Lloyd S., and Martin Renner. 2003. *Penguins*. New Haven, Conn.: Yale University Press.

De Roy, Tui. 2008. "North Pacific Survivors: The Northern Albatrosses." In *Albatross: Their World, Their Ways*, edited by Tui De Roy, Mark Jones, and Julian Fitter. Collingwood, Australia: CSIRO.

De Roy, Tui, Mark Jones, and Julian Fitter, eds. 2008. *Albatross: Their World, Their Ways*. Collingwood, Australia: CSIRO.

de Waal, Frans B. M. 2008. "Putting the Altruism Back into Altruism: The Evolution of Empathy." *Annual Review of Psychology* 59:279–300.

Decety, Jean. 2011. "The Neuroevolution of Empathy." *Annals of the New York Academy of Sciences* 1231:35–45.

Derrida, Jacques. 1993. *Aporias*. Translated by Thomas Dutoit. Stanford, Calif.: Stanford University Press.

———. 1994. *Specters of Marx*. Translated by Peggy Kamuf. New York: Routledge.

———. 1999. *Adieu to Emmanuel Levinas*. Translated by Pascale-Anne Brault and Michael Naas. Stanford, Calif.: Stanford University Press.

———. 2001. *The Work of Mourning*. Edited by Pascale-Anne Brault and Michael Naas. Chicago: University of Chicago Press.

———. 2008. "The Animal That Therefore I Am (More to Follow)." In *The Animal That Therefore I Am*. Edited by Marie-Louise Mallet. Translated by David Wills. New York: Fordham University Press.

Despret, Vinciane. 2004a. "The Body We Care For: Figures of Anthropo-zoo-genesis." *Body and Society* 10, nos. 2–3:111–34.

———. 2004b. *Our Emotional Makeup: Ethnopsychology and Selfhood*. Translated by Marjolijn de Jager. New York: Other Press.

Devinder, M. Thappa, and Kaliaperumal Karthikeyan. 2001. "Anthrax: An Overview Within the Indian Subcontinent." *International Journal of Dermatology* 40:216–22.

Diprose, Rosalyn. 2002. *Corporeal Generosity: On Giving with Nietzsche, Merleau-Ponty, and Levinas*. Albany: State University of New York Press.

Donohue, Mary J., and David G. Foley. 2007. "Remote Sensing Reveals Links Among the Endangered Hawaiian Monk Seal, Marine Debris, and El Niño." *Marine Mammal Science* 23, no. 2:468–73.

Duff, Joseph W., William A. Lishman, Clark A. Dewitt, George F. Gee, Daniel T. Sprague, and David H. Ellis. 2001. "Promoting Wildness in Sandhill Cranes Conditioned to Follow an Ultralight Aircraft." In *Proceedings of the Eighth North American Crane Workshop*, edited by David H. Ellis and Catherine H. Ellis. Seattle: North American Crane Working Group.

Ellis, David H., and George F. Gee. 2001. "Whooping Crane Egg Management: Options and Consequences." In *Proceedings of the Eighth North American Crane Workshop*, edited by David H. Ellis and Catherine H. Ellis. Seattle: North American Crane Working Group.

Ellis, David H., George F. Gee, Kent R. Clegg, Joseph W. Duff, William A. Lishman, and William J. L. Sladen. 2001. "Lessons from the Motorized Migrations." In *Proceedings of the Eighth North American Crane Workshop*, edited by David H. Ellis and Catherine H. Ellis. Seattle: North American Crane Working Group.

Ellis, David H., William J. L. Sladen, William A. Lishman, Kent R. Clegg, George F. Gee, and James C. Lewis. 2003. "Motorized Migrations: The Future or Mere Fantasy?" *BioScience* 53, no. 3:260–64.

Emery, Nathan J. 2004. "Are Corvids 'Feathered Apes'? Cognitive Evolution in Crows, Jays, Rooks and Jackdaws." In *Comparative Analysis of Minds*, edited by Shigeru Watanabe. Tokyo: Keio University Press.

Emery, Nathan J., and Nicola S. Clayton. 2004. "The Mentality of Crows: Convergent Evolution of Intelligence in Corvids and Apes." *Science* 306:1903–7.

Enright, D. J. 1983. "Editor's Note." In *The Oxford Book of Death*, edited by D. J. Enright. Oxford: Oxford University Press.

Evening News (Sydney). 1912. "Fairy Penguins. Visitors from the South. Invade Manly Beach."

Ferguson-Lees, James, and David A. Christie. 2001. *Raptors of the World*. New York: Houghton Mifflin Harcourt.

Fernandez, Patricia, David J. Anderson, Paul R. Sievert, and Kathryn P. Huyvaert. 2001. "Foraging Destinations of Three Low-latitude Albatross (*Phoebastria*) Species." *Journal of Zoology* 254:391–404.

Finkelstein, Myra E., Keith A. Grasman, Donald A. Croll, Bernie R. Tershy, Bradford S. Keitt, Walter M. Jarman, and Donald R. Smith. 2007. "Contaminant-associated Alteration of Immune Function in Black-footed Albatross (*Phoebastria nigripes*), a North Pacific Predator." *Environmental Toxicology* 26, no. 9:1896–903.

Finkelstein, Myra E., Bradford S. Keitt, Donald A. Croll, Bernie R. Tershy, Walter M. Jarman, Sue Rodriguez-Pastor, David J. Anderson, Paul R. Sievert, and Donald R. Smith. 2006. "Albatross Species Demonstrate Regional Differences in North Pacific Marine Contamination." *Ecological Applications* 16, no. 2:678–86.

Fischer, J., and D. B. Lindenmayer. 2000. "An Assessment of the Published Results of Animal Relocations." *Biological Conservation* 96, no. 1:1–11.

Foucault, Michel. 1980. *Power/Knowledge: Selected Interviews and Other Writings, 1972–1977*. Edited and translated by Colin Gordon. New York: Pantheon.

———. (1975) 1995. *Discipline and Punish: The Birth of the Prison*. Translated by Alan Sheridan. New York: Vintage.

Fraser, Orlaith N., and Thomas Bugnyar. 2010. "Do Ravens Show Consolation? Responses to Distressed Others." *PLoS ONE* 5, no. 5:1–8.

Freud, Sigmund. 1917. "Mourning and Melancholia." In *The Standard Edition of the Complete Psychological Works of Sigmund Freud*. Edited by James Strachey. Vol. 14. London: Hogarth Press.

Gee, George F., and Claire M. Mirande. 1996. "Special Techniques, Part A: Crane Artificial Insemination." In *Cranes: Their Biology, Husbandry, and Conservation*, edited by David H. Ellis, George F. Gee, and Claire M. Mirande. Washington, D.C.: Department of the Interior, National Biological Service; Baraboo, Wis.: International Crane Foundation.

Gill, Victoria. 2012. "Antarctic Moss Lives on Ancient Penguin Poo." BBC News, Nature, July 5. http://www.bbc.co.uk/nature/18704332 (accessed July 2, 2013).

Goodenough, Judith, Betty McGuire, and Elizabeth Jakob. 2010. *Perspectives on Animal Behavior*. 3rd ed. Hoboken, N.J.: Wiley.

Gregory, Murray R. 2009. "Environmental Implications of Plastic Debris in Marine Settings: Entanglement, Ingestion, Smothering, Hangers-on, Hitch-hiking and Alien Invasions." *Philosophical Transactions of the Royal Society B: Biological Sciences* 364:2013–25.

Griffiths, Tom. 2007. "The Humanities and an Environmentally Sustainable Australia." *Australian Humanities Review* 43. http://www.australianhumanitiesreview.org/

archive/Issue-December-2007/EcoHumanities/EcoGriffiths.html (accessed August 7, 2013).

Grove, Richard H. 1992. "Origins of Western Environmentalism." *Scientific American*, July, 42–47.

Grubh, Robert B. 1983. "The Status of Vultures in the Indian Subcontinent." In *Vulture Biology and Management*, edited by Sanford R. Wilbur and Jerome A. Jackson. Berkeley: University of California Press.

Gummer, Helen. 2003. *Chick Translocation as a Method of Establishing New Surface-nesting Seabird Colonies: A Review*. DOC Science Internal Series 150. Wellington, New Zealand: Department of Conservation.

Guruge, K. S., H. Tanaka, and S. Tanabe. 2001. "Concentration and Toxic Potential of Polychlorinated Biphenyl Congeners in Migratory Oceanic Birds from the North Pacific and the Southern Ocean." *Marine Environmental Research* 52, no. 3: 271–88.

Hall, Matthew. 2011. *Plants as Persons: A Philosophical Botany*. Albany: State University of New York Press.

Haraway, Donna. 1989. *Primate Visions: Gender, Race, and Nature in the World of Modern Science*. New York: Routledge.

———. 1991. "Situated Knowledges: The Science Question in Feminism and the Privilege of Partial Perspective." In *Simians, Cyborgs, and Women: The Reinvention of Nature*. New York: Routledge.

———. 1997. *Modest_Witness@Second_Millenium.FemaleMan©_Meets_OncoMouse™: Feminism and Technoscience*. New York: Routledge.

———. 2003. *The Companion Species Manifesto: Dogs, People, and Significant Otherness*. Chicago: Prickly Paradigm Press.

———. 2004. "A Manifesto for Cyborgs: Science, Technology, and Social Feminism in the 1980s." In *The Haraway Reader*. New York: Routledge.

———. 2008. *When Species Meet*. Minneapolis: University of Minnesota Press.

———. 2011. "Zoöpolis, Becoming Worldly, and Trans-species Urban Theory: For Old Cities Yet to Come." Paper presented at "Playing Cat's Cradle with Companion Species," Wellek Library Lectures, University of California, Irvine, May 5.

———. 2013. "Sowing Worlds: A Seedbag for Terraforming with Earth Others." In *Beyond the Cyborg: Adventures with Donna Haraway*, edited by Margret Grebowicz and Helen Merrick. New York: Columbia University Press.

———. Forthcoming. "Playing String Figures with Companion Species: Staying with the Trouble." In *Que savons-nous des animaux?*, edited by Vinciane Despret.

Hatley, James. 2000. *Suffering Witness: The Quandary of Responsibility After the Irreparable*. Albany: State University of New York Press.

———. 2012. "The Virtue of Temporal Discernment: Rethinking the Extent and Coherence of the Good in a Time of Mass Species Extinctions." *Environmental Philosophy* 9, no. 1:1–21.

Heidegger, Martin. 1996. *Being and Time*. Translated by Joan Stambaugh. Revised by Dennis J. Schmidt. Albany: State University of New York Press.

Heinrich, Bernd, and Thomas Bugnyar. 2007. "Just How Smart Are Ravens?" *Scientific American*, April, 64–71.

Helmreich, Stefan. 2009. *Alien Ocean: Anthropological Voyages in Microbial Seas*. Berkeley: University of California Press.

Hess, Eckhard H. 1958. "Imprinting in Animals." *Scientific American*, March, 81–90.

———. 1964. "Imprinting in Birds." *Science* 146:1128–39.

Hinchliffe, Steven, and Sarah Whatmore. 2006. "Living Cities: Towards a Politics of Conviviality." *Science as Culture* 15, no. 2:123–38.

Hird, Myra J. 2009. *The Origins of Sociable Life: Evolution After Science Studies*. New York: Palgrave Macmillan.

Horwich, Robert H., John Wood, and Ray Anderson. 1988. "Release of Sandhill Crane Chicks Hand-reared with Artificial Stimuli." In *Proceedings of the 1988 North American Crane Workshop*, edited by D. A. Wood. Tallahassee: Florida Game and Fresh Water Fish Commission.

Hosey, Geoff, Vicky Melfi, and Sheila Pankhurst. 2009. *Zoo Animals: Behaviour, Management and Welfare*. Oxford: Oxford University Press.

Houston, David C. 1983. "The Adaptive Radiation of the Griffon Vultures." In *Vulture Biology and Management*, edited by Sanford R. Wilbur and Jerome A. Jackson. Berkeley: University of California Press.

———. 2001. *Condors and Vultures*. Stillwater, Minn.: Voyageur Press.

Houston, David C., and J. E. Cooper. 1975. "The Digestive Tract of the Whiteback Griffon Vulture and Its Role in Disease Transmission Among Wild Ungulates." *Journal of Wildlife Diseases* 11:306–13.

Hughes, Janice M. 2008. *Cranes: A Natural History of a Bird in Crisis*. Richmond Hill, Ont.: Firefly Books.

Hull, Cindy L., Mark A. Hindell, Rosemary P. Gales, Ross A. Meggs, Diane I. Moyle, and Nigel P. Brothers. 1998. "The Efficacy of Translocating Little Penguins *Eudyptula minor* During an Oil Spill." *Biological Conservation* 86:393–400.

Hume, Julian P. 2006. "The History of the Dodo *Raphus cucullatus* and the Penguin of Mauritius." *Historical Biology* 18, no. 2:65–89.

Hume, Julian P., David M. Martill, and Christopher Dewdney. 2004. "Dutch Diaries and the Demise of the Dodo." *Nature*, June 10, 622.

Hunt, Gavin R. 1996. "Manufacture and Use of Hook-Tools by New Caledonian Crows." *Nature*, January 18, 249–51.

Hyrenbach, K. David, Patricia Fernandez, and David J. Anderson. 2002. "Oceanographic Habitats of Two Sympatric North Pacific Albatrosses During the Breeding Season." *Marine Ecology Progress Series* 233:283–301.

Iglesias, Teresa L., Richard McElreatha, and Gail L. Patricelli. 2012. "Western Scrub-Jay Funerals: Cacophonous Aggregations in Response to Dead Conspecifics." *Animal Behaviour* 84, no. 5:1103–11.

Immelmann, Klaus. 1972. "Sexual and Other Long-term Aspects of Imprinting in Birds and Other Species." *Advances in the Study of Behavior* 4:147–74.

Jablonka, Eva, and Marion J. Lamb. 2005. *Evolution in Four Dimensions: Genetic, Epigenetic, Behavioral, and Symbolic Variation in the History of Life*. Cambridge, Mass.: MIT Press.

Jablonski, David, and W. G. Chaloner. 1994. "Extinctions in the Fossil Record." *Philosophical Transactions of the Royal Society B: Biological Sciences* 344:11–17.

Jackson, Andrew L., Graeme D. Ruxton, and David C. Houston. 2008. "The Effect of Social Facilitation on Foraging Success in Vultures: A Modelling Study." *Biology Letters* 4:311–13.

Janzen, Daniel H., and Paul S. Martin. 1982. "Neotropical Anachronisms: The Fruits the Gomphotheres Ate." *Science* 215:19–27.

Johannesen, Edda, Lyndon Perriman, and Harald Steen. 2002. "The Effect of Breeding Success on Nest and Colony Fidelity in the Little Penguin (*Eudyptula minor*) in Otago, New Zealand." *Emu* 102:241–47.

Jones, Mark. 2008. "Perspectives: Albatrosses and Man Through the Ages." In *Albatross: Their World, Their Ways*, edited by Tui De Roy, Mark Jones, and Julian Fitter. Collingwood, Australia: CSIRO.

Jordan, Chris. 2009. "Midway: Message from the Gyre." Chris Jordan Photographic Arts. http://www.chrisjordan.com/gallery/midway (accessed July 2, 2013).

Juvik, J. O., and S. P. Juvik. 1984. "Mauna Kea and the Myth of Multiple Use: Endangered Species and Mountain Management in Hawaii." *Mountain Research and Development* 4, no. 3:191–202.

Kaufman, Leslie. 2012. "When Babies Don't Fit Plan, Question for Zoos Is, Now What?" *New York Times*, August 2.

Kearney, Richard. 2002. *On Stories*. London: Routledge.

Kingsford, R. T., J. E. Watson, C. Lundquist, O. Venter, L. Hughes, E. L. Johnston, J. Atherton, M. Gawel, D. A. Keith, B. G. Mackey, C. Morley, H. P. Possingham, B. Raynor, H. F. Recher, and K. A. Wilson. 2009. "Major Conservation Policy Issues for Biodiversity in Oceania." *Conservation Biology* 23, no. 4:834–40.

Kirksey, S. Eben. Forthcoming. "Life in the Age of Biotechnology." In *The Multispecies Salon: Gleanings from a Para-site*, edited by S. Eben Kirksey. Durham, N.C.: Duke University Press.

Kohn, Eduardo. 2007. "How Dogs Dream: Amazonian Natures and the Politics of Transspecies Engagement." *American Ethnologist* 34, no. 1:3–24.

Konig, Claus. 1983. "Interspecific and Intraspecific Competition for Food Among Old World Vultures." In *Vulture Biology and Management*, edited by Sanford R. Wilbur and Jerome A. Jackson. Berkeley: University of California Press.

Kubota, Masahisa. 1994. "A Mechanism for the Accumulation of Floating Marine Debris North of Hawaii." *Journal of Physical Oceanography* 24:1059–64.

Latimer, Joanna, and Maria Puig de la Bellacasa. 2013. "Re-Thinking the Ethical: Everyday Shifts of Care in Biogerontology." In *Ethics, Law and Society*, edited by Nicky Priaulx and Anthony Wrigley. London: Ashgate.

Leonard, David L., Jr. 2008. "Recovery Expenditures for Birds Listed Under the US Endangered Species Act: The Disparity Between Mainland and Hawaiian Taxa." *Biological Conservation* 141:2054–61.

Lestel, Dominique. 2011. "The Philosophical Stakes of Ethology for the 21st Century." Paper presented at the workshop "The History, Philosophy and Future of Ethology," Macquarie University, Sydney, February 19–21.

Lestel, Dominique, Florence Brunois, and Florence Gaunet. 2006. "Etho-ethnology and Ethno-ethology: The Coming Synthesis." *Social Science Information* 45, no. 2:155–77.

Lestel, Dominique, and Christine Rugemer. 2008. "Strategies of Life." *Research EU: The Magazine of the European Research Area*, November, 8–9.

Levine, George. 2006. *Darwin Loves You: Natural Selection and the Re-enchantment of the World*. Princeton N.J.: Princeton University Press.

Lindsey, Terence. 2008. *Albatrosses*. Collingwood, Australia: CSIRO.

Livezey, Bradley C. 1993. "An Ecomorphological Review of the Dodo (*Raphus cucullatus*) and Solitaire (*Pezophaps solitaria*), Flightless Columbiformes of the Mascarene Islands." *Journal of Zoology* 230:247–92.

Lorenz, Konrad Z. 1937. "The Companion in the Bird's World." *Auk* 54:245–73.

———. (1949) 2002. *King Solomon's Ring: New Light on Animal Ways*. Translated by Marjorie Kerr Wilson. London: Routledge.

Lorimer, Jamie. 2007. "Nonhuman Charisma." *Environment and Planning D: Society and Space* 25:911–32.

Ludwig, James P., Cheryl L. Summer, Heidi J. Auman, Vanessa Gauger, Darcy Bromley, John P. Giesy, Rosalind Rolland, and Theo Colborn. 1998. "The Roles of Organochlorine Contaminants and Fisherise Bycatch in Recent Population Changes of Black-footed and Laysan Albatrosses in the North Pacific Ocean." In *Albatross Biology and Conservation*, edited by Graham Robertson and Rosemary Gales. Chipping Norton, Australia: Surrey Beatty.

MacKinnon, John, Yvonne I. Verkuil, and Nicholas Murray. 2012. *IUCN Situation Analysis on East and Southeast Asian Intertidal Habitats, with Particular Reference to the Yellow Sea (Including the Bohai Sea)*. Gland, Switzerland: International Union for Conservation of Nature.

Malein, Flora. 2012. "Siberian Cranes Under Putin's Wings Isn't a Bad Thing." *Guardian*, September 6.

Malpas, Jeff. 1998. "Death and the Unity of a Life." In *Death and Philosophy*, edited by Jeff Malpas and Robert C. Solomon. London: Routledge.

———. 2001. "Comparing Topographies: Across Paths/Around Place: A Reply to Casey." *Philosophy & Geography* 4, no. 2:231–38.

Margulis, Lynn, and Dorian Sagan. 1995. *What Is Life?* Berkeley: University of California Press.

Markandya, Anil, Tim Taylor, Alberto Longo, M. N. Murty, S. Murty, and K. Dhavala. 2008. "Counting the Cost of Vulture Decline: An Appraisal of the Human Health and Other Benefits of Vultures in India." *Ecological Economics* 67:194–204.

Martin, Tara. 2012. "Threat of Extinction Demands Fast and Decisive Action." *The Conversation*, July 24. http://theconversation.com/threat-of-extinction-demands-fast-and-decisive-action-7985 (accessed August 8, 2013).

Marzluff, John M. 2005. *In the Company of Crows and Ravens*. Illustrated by Tony Angell. New Haven: Conn.: Yale University Press.

———. 2012. *Gifts of the Crow: How Perception, Emotion, and Thought Allow Smart Birds to Behave Like Humans*. Illustrated by Tony Angell. New York: Free Press.

Mayr, Ernst. 1996. "What Is a Species, and What Is Not?" *Philosophy of Science* 63:262–77.

———. 2001. *What Evolution Is*. New York: Basic Books.

McAllister, Molly. 2008. "Imprinting—A Case of 'Birds Gone Wrong.'" *Warbler* [newsletter of the Audubon Society of Portland, Ore.], July–August. http://audubonportland.org/about/newsletter-pdfs/julyaug (accessed July 2, 2013).

McGrath, Susan. 2007. "The Vanishing." *Smithsonian Magazine*, February.

McGrew, W. C. 1998. "Culture in Nonhuman Primates?" *Annual Review of Anthropology* 27:301–28.

Menezes, Rozario. 2008. "Rabies in India." *Canadian Medical Association Journal* 178, no. 5:564–66.

Meteyer, Carol Uphoff, Bruce A. Rideout, Martin Gilbert, H. L. Shivaprasad, and J. Lindsay Oaks. 2005. "Pathology and Proposed Pathophysiology of Diclofenac Poisoning in Free-living and Experimentally Exposed Oriental White-backed Vultures (*Gyps bengalensis*)." *Journal of Wildlife Diseases* 41, no. 4:707–16.

Millennium Ecosystem Assessment. 2005. *Ecosystems and Human Well-Being: Current State and Trends: Findings of the Condition and Trends Working Group of the Millennium Ecosystem Assessment*. Edited by Rashid Hassan, Robert Scholes, and Neville Ash. Washington, D.C.: Island Press.

Molloy, Janice, John Bennett, and Caren Schroder. 2008. "Southern Seabird Solutions Trust: Conservation Through Cooperation." In *Albatross: Their World, Their Ways*, edited by Tui De Roy, Mark Jones, and Julian Fitter. Collingwood, Australia: CSIRO.

Morton, Timothy. 2010. *The Ecological Thought*. Cambridge, Mass.: Harvard University Press.

———. 2011. "Dawn of the Hyperobjects 2." June 26. YouTube. http://www.youtube.com/watch?v=zxpPJ16D1cY (accessed July 2, 2013).

———. 2012. "Everything We Need: Scarcity, Scale, Hyperobjects." *Architectural Design* 82, no. 4:78–81.

Mosier, Andrea E., and Blair E. Witherington. 2001. "Documented Effects of Coastal Armoring Structures on Sea Turtle Nesting Behavior." In *Proceedings of the Coastal Eco-*

systems and Federal Activities Technical Training Symposium, August 20–22, Gulf Shores, Conn., U.S. Fish and Wildlife Service.

Muller-Schwarze, Dietland. 1984. *The Behavior of Penguins: Adapted to Ice and Tropics*. Albany: State University of New York Press.

Myers, Norman, and Andrew H. Knoll. 2001. "The Biotic Crisis and the Future of Evolution." *Proceedings of the National Academy of Sciences* 98, no. 10:5389–92.

Nancy, Jean-Luc. 2002. "L'Intrus." Translated by Susan Hanson. *New Centennial Review* 2, no. 3:1–14.

Naughton, Maura B., Marc D. Romano, and Tara S. Zimmerman. 2007. "A Conservation Action Plan for Black-footed Albatross (*Phoebastria nigripes*) and Laysan Albatross (*P. immutabilis*), Ver. 1.0." http://www.fws.gov/pacific/migratorybirds/pdf/Albatross%20Action%20Plan%20ver.1.0.pdf (accessed August 7, 2013).

Nixon, Rob. 2011. *Slow Violence and the Environmentalism of the Poor*. Cambridge, Mass.: Harvard University Press.

Noske, Barbara. 1989. *Humans and Other Animals: Beyond the Boundaries of Anthropology*. London: Pluto Press.

NPWS (National Parks and Wildlife Service). 2000. *Endangered Population of Little Penguins* Eudyptula minor *at Manly, Recovery Plan*. Hurstville: NSW National Parks and Wildlife Service.

———. 2002a. *Declaration of Critical Habitat for the Endangered Population of Little Penguins at Manly*. Hurstville: NSW National Parks and Wildlife Service.

———. 2002b. *Urban Wildlife Renewal: Growing Conservation in Urban Communities*. Hurstville: NSW National Parks and Wildlife Service.

Oldland, Jo, Danny Rogers, Rob Clemens, Lainie Berry, Grainne Macguire, and Ken Gosbell. 2009. *Shorebird Conservation in Australia*. Birds Australia Conservation Statement, no. 14. Carlton: Birds Australia.

Olsen, Glenn H., Jonanna A. Taylor, and George F. Gee. 1997. "Whooping Crane Mortality at Patuxent Wildlife Research Center, 1982–95." In *Proceedings of the Seventh North American Crane Workshop*, edited by Richard P. Urbanek and Dale W. Stahlecker. Grand Island, Neb.: North American Crane Working Group.

Olsen, Penny, and Leo Joseph. 2011. *Stray Feathers: Reflections on the Structure, Behaviour and Evolution of Birds*. Collingwood, Australia: CSIRO.

Oyama, Susan. 2000. *Evolution's Eye: A Systems View of the Biology–Culture Divide*. Durham, N.C.: Duke University Press.

Pain, D. J., A. A. Cunningham, P. F. Donald, J. W. Duckworth, D. C. Houston, T. Katzner, J. Parry-Jones, C. Poole, V. Prakash, P. Round, and R. Timmins. 2003. "Causes and Effects of Temporospatial Declines of Gyps Vultures in Asia." *Conservation Biology* 17, no. 3:661–71.

Palmer, Clare. 2010. *Animal Ethics in Context*. New York: Columbia University Press.

Peacock, Laurel. 2009. "*Animots* and the *Alphabête* in the Poetry of Francis Ponge." *Australian Humanities Review* 47:89–97.

Pichel, William G., James H. Churnside, Timothy S. Veenstra, David G. Foley, Karen S. Friedman, Russell E. Brainard, Jeremy B. Nicoll, Quanan Zheng, and Pablo Clemente-Colon. 2007. "Marine Debris Collects Within the North Pacific Subtropical Convergence Zone." *Marine Pollution Bulletin* 54:1207–11.

Pika, Simone, and Thomas Bugnyar. 2011. "The Use of Referential Gestures in Ravens (*Corvus corax*) in the Wild." *Nature Communications* 2, no. 560:1–5.

Plumwood, Val. 2002. *Environmental Culture: The Ecological Crisis of Reason*. London: Routledge.

———. 2003. "Animals and Ecology: Towards a Better Integration." Australian National University. http://hdl.handle.net/1885/41767 (accessed July 2 2013).

———. 2007. "Human Exceptionalism and the Limitations of Animals: A Review of Raimond Gaita's *The Philosopher's Dog*." *Australian Humanities Review* 42.

———. 2008a. "Shadow Places and the Politics of Dwelling." *Australian Humanities Review* 44:139–50.

———. 2008b. "Tasteless: Towards a Food-based Approach to Death." *Environmental Values* 17:323–30.

———. 2009. "Nature in the Active Voice." *Australian Humanities Review* 46:113–29.

———. 2011. "'Babe': The Tale of the Speaking Meat: Part I." *Australian Humanities Review* 51:205–7.

Podolsky, Richard H. 1990. "Effectiveness of Social Stimuli in Attracting Laysan Albatross to New Potential Nesting Sites." *Auk* 107:119–25.

Poole, Joyce. 1996. *Coming of Age with Elephants: A Memoir*. London: Hodder and Stoughton.

Prakash, V. 1999. "Status of Vultures in Keoladeo National Park, Bharatpur, Rajasthan, with Special Reference to Population Crash in *Gyps* Species." *Journal of the Bombay Natural History Society* 96:365–78.

Prakash, V., R. E. Green, D. J. Pain, S. P. Ranade, S. Saravanan, N. Prakash, R. Venkitachalam, R. Cuthbert, A. R. Rahmani, and A. A. Cunningham. 2007. "Recent Changes in Populations of Resident *Gyps* Vultures in India." *Journal of the Bombay Natural History Society* 104, no. 2:129–35.

Primack, Richard. 1993. *Essentials of Conservation Biology*. Sunderland, Mass.: Sinaur.

Puig de la Bellacasa, Maria. 2012. "'Nothing Comes Without Its World': Thinking with Care." *Sociological Review* 60, no. 2:197–216.

Quammen, David. 1996. *The Song of the Dodo: Island Biogeography in an Age of Extinctions*. New York: Scribner.

Raup, David M., and J. John Sepkoski. 1982. "Mass Extinctions in the Marine Fossil Record." *Science* 215:1501–3.

Read, Peter. 1996. *Returning to Nothing: The Meaning of Lost Places*. Cambridge: Cambridge University Press.

Reilly, P. N., and J. M. Cullen. 1981. "The Little Penguin *Eudyptula minor* in Victoria, II: Breeding." *Emu* 81, no. 1:1–19.

Reinert, Hugo. 2007. "The Pertinence of Sacrifice—Some Notes on Larry the Luckiest Lamb." *Borderlands* 6, no. 3:1–32.

Restani, Marco, and John M. Marzluff. 2002. "Funding Extinction? Biological Needs and Political Realities in the Allocation of Resources to Endangered Species Recovery." *Bio-Science* 52, no. 2:169–77.

Ricciardi, Alessia. 2003. *The Ends of Mourning: Psychoanalysis, Literature, Film*. Stanford, Calif.: Stanford University Press.

Rice, Dale W., and Karl W. Kenyon. 1962. "Breeding Cycles and Behavior of Laysan and Black-footed Albatrosses." *Auk* 79, no. 4:517–67.

Rich, Pat V. 1983. "The Fossil Record of the Vultures: A World Perspective." In *Vulture Biology and Management*, edited by Sanford R. Wilbur and Jerome L. Jackson. Berkeley: University of California Press.

Ricoeur, Paul. 2007. "On Stories and Mourning." In *Traversing the Imaginary: Richard Kearney and the Postmodern Challenge*, edited by Peter Gratton and John Panteleimon Manoussakis. Evaston, Ill.: Northwestern University Press.

Riegel, Christian. 2003. *Writing Grief: Margaret Laurence and the Work of Mourning*. Winnipeg: University of Manitoba Press.

Robbins, Paul. 1998. "Shrines and Butchers: Animals as Deities, Capital, and Meat in Contemporary North India." In *Animal Geographies: Place, Politics, and Identity in the Nature–Culture Borderlands*, edited by Jennifer Wolch and Jody Emel. London: Verso.

Rogers, T., and C. Knight. 2006. "Burrow and Mate Fidelity in the Little Penguin *Eudyptula minor* at Lion Island, New South Wales, Australia." *Ibis* 148:801–6.

Rolston, Holmes, III. 1998. "Down to Earth: Persons in Place in Natural History." In *Philosophy and Geography III: Philosophies of Place*, edited by Andrew Light and Jonathan M. Smith. Lanham, Md.: Rowman & Littlefield.

———. 1999. "Respect for Life: Counting What Singer Finds of No Account." In *Singer and His Critics*, edited by Dale Jamieson. Oxford: Blackwell.

Rose, Deborah Bird. 2006. "What If the Angel of History Were a Dog?" *Cultural Studies Review* 12, no. 1: 67–78.

———. 2008. "Judas Work: Four Modes of Sorrow." *Environmental Philosophy* 5, no. 2:51–66.

———. 2012a. "In the Shadow of All This Death." Paper presented at the conference "Animal Death," University of Sydney, June 12–13. Extinction Studies Working Group. http://extinctionstudies.org/communications/shadow_of_all_this_death (accessed July 2, 2013).

———. 2012b. "Multispecies Knots of Ethical Time." *Environmental Philosophy* 9, no. 1:127–40.

Rose, Deborah Bird, and Thom van Dooren, eds. 2011. "Unloved Others: Death of the Disregarded in the Time of Extinctions." Special issue, *Australian Humanities Review* 50.

Ruxton, Graeme D., and David C. Houston. 2004. "Obligate Vertebrate Scavengers Must Be Large Soaring Fliers." *Journal of Theoretical Biology* 228:431–36.

Safina, Carl. 2007. "Wings of the Albatross." *National Geographic,* December.

———. 2008. Introduction. In *Albatross: Their World, Their Ways,* edited by Tui De Roy, Mark Jones, and Julian Fitter. Collingwood, Australia: CSIRO.

Sagan, Dorion. 2010. "Introduction: Umwelt After Uexküll." In *A Foray into the Worlds of Animals and Humans, with A Theory of Meaning,* by Jakob von Uexküll. Translated by Joseph D. O'Neil. Minneapolis: University of Minnesota Press.

Scarponi, Antonio. 2012. "The Seventh Continent—Musings on The Plastic Garbage Project." August 27. Domus. http://www.domusweb.it/en/design/the-seventh-continent-musings-on-the-plastic-garbage-project (accessed July 2, 2013).

Schuz, Ernst, and Claus Konig. 1983. "Old World Vultures and Man." In *Vulture Biology and Management,* edited by Sanford R. Wilbur and Jerome A. Jackson. Berkeley: University of California Press.

Seed, Amanda M., Nicola S. Clayton, and Nathan J. Emery. 2007. "Postconflict Third-Party Affiliation in Rooks, *Corvus frugilegus.*" *Current Biology* 17:152–58.

Seed, Amanda, Nathan Emery, and Nicola Clayton. 2009. "Intelligence in Corvids and Apes: A Case of Convergent Evolution?" *Ethology* 115:401–20.

Serventy, D. L., B. M. Gunn, I. J. Skira, J. S. Bradley, and R. D. Wooller. 1989. "Fledgling Translocation and Philopatry in a Seabird." *Oecologia* 81:428–29.

Shaffer, Scott. 2008. "Albatross Flight Performance and Energetics." In *Albatross: Their World, Their Ways,* edited by Tui De Roy, Mark Jones, and Julian Fitter. Collingwood, Australia: CSIRO.

Sileo, Louis, Paul R. Sievert, and Michael D. Samuel. 1990. "Causes of Mortality of Albatross Chicks at Midway Atoll." *Journal of Wildlife Diseases* 26, no. 3:329–38.

Singh, Jyotsna. 2003. "India Targets Cow Slaughter." BBC News, August 11. http://news.bbc.co.uk/2/hi/south_asia/2945020.stm (accessed July 2, 2013).

Slagsvold, Tore, Bo T. Hansen, Lars E. Johannessen, and Jan T. Lifjeld. 2002. "Mate Choice and Imprinting in Birds Studied by Cross-fostering in the Wild." *Proceedings of the Royal Society B: Biological Sciences* 269:1449–55.

Sluckin, Wladyslaw. 1964. *Imprinting and Early Learning.* London: Methuen.

Smith, Mick. 2001. "Environmental Anamnesis: Walter Benjamin and the Ethics of Extinction." *Environmental Ethics* 23, no. 3:359–76.

———. 2011. "Dis(appearance): Earth, Ethics and Apparently (In)Significant Others." In "Unloved Others: Death of the Disregarded in the Time of Extinctions," edited by Deborah Bird Rose and Thom van Dooren, special issue, *Australian Humanities Review* 50:23–44.

Snyder, Noel F. R., Scott R. Derrickson, Steven R. Beissinger, James W. Wiley, Thomas B. Smith, William D. Toone, and Brian Miller. 1996. "Limitations of Captive Breeding in Endangered Species Recovery." *Conservation Biology* 10, no. 2:338–48.

Star, Susan Leigh, and Anselm Strauss. 1999. "Layers of Silence, Arenas of Voice: The Ecology of Visible and Invisible Work." *Computer Supported Cooperative Work* 8:9–30.

Steadman, David W. 1995. "Prehistoric Extinctions of Pacific Island Birds: Biodiversity Meets Zooarchaeology." *Science* 267:1123–31.

———. 2006. *Extinction and Biogeography of Tropical Pacific Birds.* Chicago: University of Chicago Press.

Stearns, Beverly Peterson, and Stephen C. Stearns. 1999. *Watching, from the Edge of Extinction.* New Haven, Conn.: Yale University Press.

Stewart, Will. 2012. "Now It's Putin the Bird Man: Latest Animal Stunt See Russian President Take to Skies in Micro Glider as 'Chief Crane.'" *Daily Mail,* September 5. http://www.dailymail.co.uk/news/article-2198963/Now-Putin-bird-man-Latest-animal-stunt-sees-Russian-president-skies-micro-glider-chief-crane.html (accessed July 2, 2013)

Strauss, Jonathan. 2000. "After Death." *Diacritics* 30, no. 3:90–104.

Stuart, Simon N., Michael Hoffmann, J. S. Chanson, N. A. Cox, R .J. Berridge, P. Ramani, and B. E. Young. 2008. *Threatened Amphibians of the World.* Barcelona: Lynx Edicions.

Subramanian, Meera. 2008. "Towering Silence." *Science & Spirit,* May–June, 34–38.

Sullivan, Ben. 2008. "The Albatross Task Force: A Sea Change for Seabirds." In *Albatross: Their World, Their Ways,* edited by Tui De Roy, Mark Jones, and Julian Fitter. Collingwood, Australia: CSIRO.

Sunday Telegraph. 1954. "300 Penguins Shot at North Head." August 22, 3.

Sun-Herald. 1954. "Hoodlums Kill 30 Penguins." October 24, 7.

Swan, Gerry E., Richard Cuthbert, Miguel Quevedo, Rhys E. Green, Deborah J Pain, Paul Bartels, Andrew A. Cunningham, Neil Duncan, Andrew A. Meharg, Lindsay Oaks, Jemima Parry-Jones, Susanne Shultz, Mark A. Taggart, Gerhard Verdoorn, and Kerri Wolter. 2006. "Toxicity of Diclofenac to *Gyps* Vultures." *Biology Letters* 2:279–82.

Swan, Gerry, Vinasan Naidoo, Richard Cuthbert, Rhys E. Green, Deborah J. Pain, Devendra Swarup, Vibhu Prakash, Mark Taggart, Lizette Bekker, Devojit Das, Jorg Diekmann, Maria Diekmann, Elmarié Killian, Andy Meharg, Ramesh Chandra Patra, Mohini Saini, and Kerri Wolter. 2006. "Removing the Threat of Diclofenac to Critically Endangered Asian Vultures." *PLoS Biology* 4, no. 3:0395–402.

Swengel, Scott R., George W. Archibald, David H. Ellis, and Dwight G. Smith. 1996. "Behavior Management." In *Cranes: Their Biology, Husbandry, and Conservation,* edited by David H. Ellis, George F. Gee, and Claire M. Mirande. Washington, D.C.: Department of the Interior, National Biological Service; Baraboo, Wis.: International Crane Foundation.

Sydney Morning Herald. 1931. "Crested Penguin." June 15, 8.

———. 1936. "Penguins: Minister Warns Public." March 31, 9.

———. 1948. "'Fairies' in the Harbour." March 26, 2.

Taylor, Alex H., Gavin R. Hunt, Jennifer C. Holzhaider, and Russell D. Gray. 2007. "Spontaneous Metatool Use by New Caledonian Crows." *Current Biology* 17:1504–7.

Temple, Stanley A. 1977. "Plant–Animal Mutualism: Coevolution with Dodo Leads to Near Extinction of Plant." *Science* 197:885–86.

ten Cate, Carel, and Dave R. Vos. 1999. "Sexual Imprinting and Evolutionary Processes in Birds: A Reassessment." *Advances in the Study of Behavior* 28:1–31.

Thomson, Melanie S. 2007. "Placing the Wild in the City: 'Thinking with' Melbourne's Bats." *Society and Animals* 15:79–95.

Thornton, Joe. 2000. *Pandora's Poison: Chlorine, Health, and a New Environmental Strategy*. Cambridge, Mass.: MIT Press.

Tsing, Anna Lowenhaupt. 2012. "Unruly Edges: Mushrooms as Companion Species." *Environmental Humanities* 1:141–54.

———. Forthcoming. "Blasted Landscapes, and the Gentle Art of Mushroom Picking." In *The Multispecies Salon: Gleanings from a Para-site*, edited by S. Eben Kirksey. Durham, N.C.: Duke University Press.

U.S. Fish and Wildlife Service (USFWS). 2007. *International Recovery Plan for the Whooping Crane* (Grus americana), *Third Revision*. Albuquerque: U.S. Fish and Wildlife Service.

———. 2009. *Revised Recovery Plan for the 'Alalā* (Corvus hawaiiensis). Portland, Ore.: U.S. Fish and Wildlife Service.

———. 2011. "Endangered and Threatened Wildlife and Plants: Establishment of a Nonessential Experimental Population of Endangered Whooping Cranes in Southwestern Louisiana." *Federal Register* 76, no. 23:6066–82.

———. 2012. "Midway Atoll National Wildlife Refuge—About Us." April 30. http://www.fws.gov/midway/aboutus.html (accessed July 2, 2013).

van Dooren, Thom. 2010. "Pain of Extinction: The Death of a Vulture." *Cultural Studies Review* 16, no. 2:271–89.

———. 2011a. "Invasive Species in Penguin Worlds: An Ethical Taxonomy of Killing for Conservation." *Conservation and Society* 9, no. 4:286–98.

———. 2011b. *Vulture*. London: Reaktion Books.

van Dooren, Thom, and Deborah Bird Rose. 2012. "Storied-places in a Multispecies City." *Humanimalia* 3, no. 2:1–27.

Vijaikumar, M., M. Thappa Devinder, and K. Karthikeyan. 2002. "Cutaneous Anthrax: An Endemic Outbreak in South India." *Journal of Tropical Pediatrics* 48, no. 4:225–26.

von Uexküll, Jakob. (1934, 1940) 2010. *A Foray into the Worlds of Animals and Humans, with A Theory of Meaning*. Translated by Joseph D. O'Neil. Minneapolis: University of Minnesota Press.

Walters, Mark Jerome. 2006. *Seeking the Sacred Raven: Politics and Extinction on a Hawaiian Island*. Washington, D.C.: Island Press.

Warkentin, Tracy. 2010. "Interspecies Etiquette: An Ethics of Paying Attention to Animals." *Ethics and the Environment* 15, no. 1:101–21.

Wellington, Marianne, Ann Burke, Jane M. Nicholich, and Kathleen O'Malley. 1996. "Chick Rearing." In *Cranes: Their Biology, Husbandry, and Conservation*, edited by David H. Ellis, George F. Gee, and Claire M. Mirande. Washington, D.C.: Department of the Interior, National Biological Service; Baraboo, Wis.: International Crane Foundation.

West, Meredith J., and Andrew P. King. 1987. "Settling Nature and Nurture into an Ontogenetic Niche." *Developmental Psychobiology* 20, no. 5:549–62.

Wexler, Rebecca. 2008. "Onward, Christian Penguins: Wildlife Film and the Image of Scientific Authority." *Studies in History and Philosophy of Biological and Biomedical Sciences* 39:273–79.

Wilcox, Bruce A. 1988. "Tropical Deforestation and Extinction." In *IUCN Red List of Threatened Animals*, edited by International Union for Conservation of Nature and Natural Resources. Gland, Switzerland: International Union for Conservation of Nature.

Wilkins, John S. 2009. *Species: A History of the Idea*. Berkeley: University of California Press.

Williams, Alan. 1997. "Zoroastrianism and the Body." In *Religion and the Body*, edited by Sarah Coakley. Cambridge: Cambridge University Press.

Wolch, Jennifer. 2002. "Anima urbis." *Progress in Human Geography* 26, no. 6:721–42.

Wolfe, Cary. 2009. *What Is Posthumanism?* Minneapolis: University of Minnesota Press.

Wylie, Dan. 2010. "Minding Elephants and the Rhetorics of Destruction." *Australian Literary Studies* 25, no. 2:72–87.

Wynne, Clive D. L. 2002. *Animal Cognition: The Mental Lives of Animals*. New York: Palgrave Macmillan.

Young, Lindsay C., Brenda J. Zaun, and Eric A. VanderWerf. 2008. "Successful Same-Sex Pairing in Laysan Albatross." *Biology Letters* 4:323–25.

INDEX

Numbers in italics refer to pages on which illustrations appear.